# NO TIME
# FOR HISTORY

# NO TIME
# FOR HISTORY

*a pioneer story*

by ARIE L. ELIAV

translated from Hebrew by Dov Chaikin

SABRA BOOKS • NEW YORK

SBN 87631-032-3
Library of Congress Catalog Card Number 72-124128

*To my daughter Ofra
and the residents
of the Lakhish Region*

# CONTENTS

# CHAPTER I

## FIRST DAYS' WORK WITH ESHKOL

I feel the burden of the years. Before the winds of time scatter the memories like so much dust I wish to recount the days of Lakhish and the massive settlement program of the early years of our State.

Who molded me? There were two such men, both of the founding generation: Shaul Avigur, my commander during illegal immigration days, and Levi Eshkol who inherited me from Shaul after the War of Independence.

My first real meeting with Eshkol sticks in my mind to this day. It was in 1948, during the war. Eshkol whom Ben-Gurion had appointed a sort of ad hoc Deputy Minister of Defense was meeting representatives of all branches of the fledgling Defense Forces to discuss their budgets. I recall how he cut down to the size of the young State's resources the ambitions of the commanders of the various branches. All this he did with tremendous humor, which was so effective that the injured party never resented Eshkol's sharp "axe."

At the beginning of 1949, the war over, I was discharged from the Israel Defense Forces. For over ten years I had served — with others of my generation — in the *Haganah*, the British Army, *Aliyah Bet* and the Israel Defense Forces.

I asked myself: what shall I do as a civilian? I recalled my studies at the Faculty of Agriculture on Mount Scopus. I went to see Shaul Avigur, my mentor and commanding officer in the underground and illegal immigration days, and asked his advice. Shaul told me that Eshkol had recently been appointed head of the Jewish Agency's Settlement Department, and was looking for an assistant.

With Shaul's recommendation in hand, I sought an interview with Eshkol.

In those days, the mass immigration waves were already pouring into Israel. Eshkol was up to his ears in work. Many people crowded into his office.

When I was finally admitted to Eshkol, after a lengthy wait, we exchanged only a few words.

"Right," said Eshkol, "let's start from the end. When are you prepared to start work?"

I asked him innocently, from which date he wanted me to start work.

"From which date? From yesterday," was his reply.

I explained that I had not yet received my final discharge papers, and if he had no objections — I would work in uniform for a short while. Eshkol was agreeable, and then he asked : "And what's your family status ? Are you married?"

"Yes," I answered, "married and the father of a small baby."

"Oy," Eshkol groaned. "And your parents, are they alive?"

"My old mother is alive, and we live with her," I answered.

"No good, no good," Eshkol said, a mischievous glint in his eyes and a slight smile on his face.

"What's no good?" I wondered.

"You don't meet my requirements," Eshkol answered. "I had decided that the person to help me in this hard task

I'm undertaking — the settlement of Jews on the land —
must enjoy a special family status. He must be a bachelor
and an orphan. Now you come and tell me of a mother,
a wife and a baby... When will you have time for them?
When will you see them?"

That was how I started working.

The first days were days of confusion. I had never done
civilian work; I had never been anyone's "assistant." Until
then I had received and issued orders.

Here everything was different.

On the first day I settled down peacefully in the room
allocated to me, facing Eshkol's room. Through the doorway I could see a long line of people: men of the settlements
and farms, all kinds of experts, party hacks, and just plain
good Jews with advice on how to "save" the Jewish People.

Nobody asked for me and no one came in to see me the
entire day.

At the end of the day I went to Eshkol's room and asked
politely: "Can I help you in anything?"

Eshkol, exhausted from the day's toil, looked at me
through tired and angry eyes: "What do you mean, anything?
What's going to happen to all the load that's been piled on
me today?"

"What load?" I asked.

"What load, you ask? Here, for instance, we've got the
matter of the Harzfeld plan concerning the settlement of new
kibbutzim in the Negev..."

"What is Harzfeld's plan?" I asked.

"Now look, he sat here for two hours and lectured me
on it; why didn't you come in and listen?"

"I didn't come in because you didn't call me!"

"So what do you want," Eshkol asked, "I should now sit
and tell you everything I've been through all day? It would
take another day. I'm busy now and must run."

On the morrow things were much the same. Eshkol sank under the stream of people. I sat in my room, dreary and dumbfounded.

In the evening, when Eshkol got down to explaining things, it transpired that most of his "clients" were unfamiliar to me. Again and again I had to ask who so-and-so was and who was this one or that one.

"So," stormed Eshkol, "now I should tell you the biography of the people of the Second and Third *Aliyah* in one go?"

The third day, I walked into Eshkol's room and said to him: "There's no point in carrying on like this. It's a wasted effort. I'm no use to you. I'm redundant. I suggest that from tomorrow I move into your room, and sit together with you and your 'clients' from morning till night. Only thus, maybe, will I be able to penetrate into the thick of things and help you."

Eshkol accepted my suggestion willingly, and from then on the way was opened to close cooperation between us, which was to last for many years. Our work brought us closer from day to day, the ties eventually becoming the strongest and firmest possible. Admittedly, in the beginning my presence seemed strange to the people calling on Eshkol but gradually they got used to me.

Aside from the many people passing through Eshkol's room, we were flooded with letters. I tried to answer every letter and attend to the matter raised. I devoted many hours to these letters and to answering them. Some answers I would sign myself, others I would get Eshkol to sign.

One day Eshkol said to me: "I see you're putting what's left of your energy into these letters, answering every V.I.P. and every *nudnik* writing to us. Here's a suggestion how to simplify the process and save energy, time and stamps. For a whole month don't touch the mail. Take all incoming mail, without exception, even if the envelope has on it 'personal,' 'urgent,' 'express,' 'immediate,' 'addressee only' and other such trimmings. Put all the letters in a big drawer in your desk. Do not open a single one. During this month we will, of course, attend to our affairs: we shall see lots of people, arrange many matters, tour many places and make many decisions.

"At the end of the month you'll open all the accumulated letters, and you will then see that the mail can be divided into two categories: letters containing really urgent and important matters, and letters from *nudniks* and cranks.

"You will then also realize that matters within the first category have already worked themselves out or are receiving our urgent attention. For that's the nature of important subjects, that they reach us in ten different ways — whether through personal interviews (after all, you don't refuse anyone asking us for an interview), at meetings, via the Press, and so forth.

"Dealing with letters of the second category is a waste of time. Ergo: why bother to answer mail at all?"

He then added with a smile: "I don't object, of course, to your answering every single letter, but, for Heaven's sake, don't sacrifice what little time you have at your disposal. After all, I had no success with you — you're neither bachelor nor orphan. Attend to your family duties."

CHAPTER II

## "CABOOSES"

During my first year of work, 1949, and the following year, the waves of immigration arrived, swamping little Israel and its Jewish population. It cannot be said that we were prepared for the absorption of the hundreds of thousands of our brethren, who arrived by sea and air, from West and East. It was then that the myriads of survivors from Eastern Europe arrived, broken in body and spirit; those were the days of "Operation Magic Carpet," when the Yemenite exiles — men, women and children — landed in their thousands; Bulgaria was emptied of her deep-rooted Jews; the Jews of Greece and Yugoslavia arrived; the flow commenced of Iraqi Jews, and the Jews of the Libyan and Moroccan Diaspora.

We had no organized means of absorption. The work of bringing in, absorbing and settling the Jews, was performed through constant improvisation, under ever-growing strain. We needed every grain of counsel and resourcefulness to withstand the pressure.

And the most difficult problem, indeed the first we encountered, was that of housing for the immigrants. Within a very short while, in what seemed like no time at all, the structures abandoned by the Arabs in the towns were filled. Not all of them were fit for habitation. Many were ramshackle and a danger to the occupants. We filled the British Army camps which had fallen into our hands. We went over to tents — and then the tents ran out. We asked

14

for, and received, a few tents from the Israel Defense Forces. These were also filled, and we were obliged to open workshops to patch old tents we could lay our hands on here and there. Very soon, an urgent appeal was sent by the Jewish Agency to all its emissaries throughout the world: "Let us have tents. Let us have any means of accommodation which can be loaded onto ships and brought to Israel."

At that time there was an emissary of the Settlement Department in the United States, a fellow by the name of Giladi, a member of Kibbutz Deganya Bet, an agricultural equipment expert, who had been sent to procure new equipment for the new kibbutzim and *moshavim*.

One day we received an urgent cable from him from New York, as follows :

"Have acquired hundred cabooses good condition dirt cheap. Have option further thousand. Preparing send first hundred earliest ship for Israel."

I turned the cable this way and that, scratched my forehead, but my Hebrew and foreign lexicon would not come up with anything resembling "cabooses."

I showed Eshkol the cable, perhaps he knew what "cabooses" were.

No, he did not know; he had never heard of "cabooses."

We called in the Technical Department's engineers, Yehuda Rabinovitz and Alexander Bassin, experienced veterans. Neither of them knew what a caboose was. We had no choice but to despatch a cable to Giladi asking him what, in Heaven's name, cabooses were. To be on the safe side, Eshkol added: "Don't take up option on thousand before replying."

A few days later we received a vague reply, reading as follows:

"Cabooses are good wooden structures, mobile, on wheels."
Giladi further added his opinion that every one of them

could house one family comfortably.

Eshkol asked me to cable Giladi, as follows :

"Don't purchase additional cabooses. Don't send all hundred. Send ten next ship. Will check here and let you know further."

Giladi advised that he was loading ten cabooses on so-and-so ship, due to arrive in Haifa port on such-and-such a date. In time, we received a bill of lading in the mail for ten cabooses. The weight noted in the bill had us scared. It said that the weight of each caboose was a few tons.

On the appointed day I set out for Haifa port with two of our engineers.

"Your cargo is on board," said the customs officials, "and will soon be unloaded."

We stood next to the ship and awaited developments. The crane groped around in the ship's hold and started hauling out large wooden crates, some rectangular, some square, some oblong, and some even round. A large stack piled up on the wharf. The engineers took a close look. Their eyes opened up in surprise. "This looks like a wall," said one. "And here's something made of metal," said the other. "Hold on a minute, we'll open the packing." He looked in and withdrew : "There's a stove and chimney inside..." He was obviously perplexed. The two approached a flat, round case, peeping into that as well : "Look, look, it's a large wheel !" "A wheel !" I called out. "What's a wheel doing here ?"

Meanwhile, a curious crowd gathered around us — porters, customs officers and spectators. "What's this?" "Where's it from?" "What's it for?" We decided to decamp with the whole caboodle. We carted the cabooses on a freight train to our warehouses at Tzrifim. There, in among the eucalyptus groves, we could seclude ourselves with our "treasure" and examine it closely.

The following day, we brought in a gang of laborers from

Ramleh to unpack. It was not long before the cabooses stood revealed in all their glory — good, thick wooden cabins, with a legend in English: "North-Central Railroad Company."

From other crates we removed large, heavy train wheels, chairs, tables and beds attachable with screws to the wooden floor and walls, a large stove to be fixed in the center of the carriage with a chimney leading from it to the ceiling, as well as carriage windows and doors.

Our engineers — who, for their sins, had not studied railway engineering, contenting themselves with civil engineering — spent a hard day's labor assembling one model carriage. A large and impressive wheeled carriage, with space for five or six persons.

We summoned the Israel Railways' Engineer from Lod, to give his opinion of this "dinosaur." He arrived, marveled at the caboose, examined it very carefully, and told us that the "caboose" was a railway carriage which had once been employed principally in the United States, and had for some time fallen into disuse. When in use, it had served as living quarters for the railmen and train guards, who had to travel long hours and days over America's extensive railroads. In it they slept, ate, made tea or coffee and warmed themselves. Since they ceased using this carriage, the railroad companies were seeking buyers for the "bargain." That, presumably, was how they reached Giladi.

We reported to Eshkol, who was sufficiently curious to want to see a caboose with his own eyes. He came out to Tzrifim with us for a glimpse at the marvel.

That very day an urgent cable was despatched to Giladi: "Intentions good but no use found for cabooses. Sell remaining ninety units whatever price."

The cabooses remained in Tzrifim, abandoned and forgotten. Not for all time, though.

Years later, when we set about the establishment of the

development and minerals companies in the Negev, Eshkol remembered the cabooses, lying unwanted. Approached by the person in charge of development of the Dead Sea Potash Works at Sdom, with the usual request for a budget, Eshkol said to him : "Friend, you will receive so many hundreds of thousands of pounds, plus a caboose. Ask no questions and don't try to understand. Here's a note to the storeman at Tzrifim. Take one caboose from him and set it up in Sdom. It's an excellent thing and will prove useful to you."

Eshkol thus distributed gifts of cabooses to the Phosphate Works at Oron and to the Timna Copper Mines in the Arava. What they did with them — I neither know nor care.

The problem of immigrant housing, however, was still unsolved. One day Eshkol summoned a concilium of the engineers and architects for the umpteenth time, of those engaged in immigration, absorption and settlement.

The immigration people provided forecasts of the number of immigrants due in the coming months. The absorption official and the engineers submitted a report on the position of accommodation for the immigrants. The gap between both figures gave us no peace. Turning to Bassin, the senior engineer, Eshkol asked :

"What have we got in the warehouses ? What building materials from which we can quickly assemble any kind of structures ?"

The engineer consulted his lists.

"We don't have enough timber to put up provisional huts, nor any canvas or tarpaulin for the sewing of tents. We have a large quantity of corrugated metal sheets, ordered for the construction of roofs for cowsheds and chicken-coops in the agricultural settlements. But what can you do with corrugated metal sheets ?"

"Hold on," Eshkol interrupted him, "maybe you'll use it to construct something resembling a hut ? A hut made of corrugated metal sheets ?"

"But we have no timber for walls," the engineers interjected.

"So maybe we'll make walls of metal sheets as well ?"

The engineers held a consultation, made sketches, exchanged views. Finally, Bassin submitted to us the design of a box-like structure, made up of a wooden frame, but with corrugated metal sheets for walls and roof.

These "tin cans," 15 square meters in area and capable of housing an immigrant family, could be produced in large quantities within a short while. And that was how the tin hut was born.

Tin huts in their thousands started covering the land. Any number of deficiencies were to be found in them, they froze in winter and were baking hot in summer. But they did provide a temporary solution, for all that this temporariness lasted a number of years.

When the corrugated metal sheets also ran out, the canvas hut was invented. The corrugated sheet metal walls were replaced by partitions made up of remnants of tarpaulin, which could no longer be used for tents.

Camps of shacks, and of huts from tin and canvas sprouted like mushrooms near town and village. In the center of each camp a number of double tin huts or shacks were put up, serving as stores, Jewish Agency offices, synagogues, clinics, schools, kindergartens and clubhalls. Faucets and primitive sanitary installations, made of metal sheets or wood, were put up at intervals. People, dust and constant tumult filled the camps. Life in them was congested, dense and stifling. These were the *ma'abarot*.

Nowadays, there is almost no sign left of the dozens of *ma'abarot* which dotted the countryside. A lone surviving tin hut, converted into an outhouse, may still be observed here and there in the yard of one of the urban housing projects. The day will come when those who spent their childhood and youth in the *ma'abarot,* who grew up in the tin huts and the canvas huts, will be as proud of that period as were the early Hadera pioneers of the *Khan* in which they lived and in which they suffered from malaria, or the Jordan Valley pioneers of the first mud huts at *Um Junni.*

"What do you know of hardship?" a fellow will ask his children. When I came to Israel with my parents, where do you think they put us? In a tin hut. What, children? You don't know what a tin hut is? Well, let me tell you..."

## CHAPTER III

# "THE SIEVE"

## *An Eshkol Miscellany*

We used to come to Eshkol in those days and pour our hearts out to him: the *ma'abarot* were filling up, there were no tents, there were demonstrations of clans and families — not omitting the aged, the women and babes in arms. What was to be done?

Eshkol would say to us: "Kinderlach Children" (this was the way he would often address his peers and elders) "our life resembles a large sieve. In this sieve, without order or plan, there are stones — large, small and medium — and pebbles, sand, silt and floating dust particles. The sieve sinks under the load of stones, and moves slowly to and fro, to and fro, steadily and constantly. Never, it would seem, will there be any order in this muddle. No stone will find its aperture, and the sieve is doomed to break and come to a halt.

"But you, children, what are you to do? Just one thing: Keep the sieve moving and never let go! Perspiration will cover your bodies and drip into your eyes, your muscles will ache and your ears echo with the mighty sound and tumult. Eventually, though, the large stones will find their large apertures, the medium-sized stones their medium-sized

21

apertures, and the small stones their small apertures; the sand will settle and the dust be borne away by the wind; and you will see before you, at the end of a day's labor, fine piles of good stones, of which roads are paved and houses built.

"Just don't let go of the sieve, not even for one minute."

## *"Opinion"*

During Eshkol's tenure as Jewish Agency Treasurer, and even more so when he became Minister of Finance, the custom of "consulting necromancers and soothsayers" became prevalent — that was when economists and economic advisers came into fashion. In those days, in the early 'fifties, Prof. Patinkin's "boys" had not yet entered the picture. The economists who still prevailed then were graduates of German and other Central European schools; elderly "Herr Doktors," with degrees in political economy, a profession much favored by wealthy Jewish parents in the "good old days" of the Emperor Franz Joseph.

Eshkol also got caught up in the spirit of the times, although his acute common sense told him that many of these advisers were already old-fashioned. In that early period of the sages of the political economy school, he displayed praiseworthy patience toward the long and important-sounding dissertations emanating from their desks. But when he came across the totally new image of the young Israeli economist, Eshkol grabbed these men with both hands and manned the various departments of his Ministry with them.

He had absolutely no objection to a "scientific grounding" being given to our work in immigrant settlement, water resources development and all the rest. I recall the times when a new idea or new approach popped into our minds, and we would put it on paper. I would pass the draft on to our economic adviser of the day, attached to a letter signed by Eshkol ending, usually, with the note "please let us have your opinion."

Whenever such a letter contained an idea which Eshkol particularly liked, he would insert the word "positive" into the note "please let us have your opinion."

And, wonder of wonders, the opinions we received were always learned, weighty, measured, scientific — and, of course, positive...

## *"Bim-Bam-Bam"*

I covered thousands of miles with Eshkol on Israel's highways and byways. The car would usually be full of people — senior department staff, settlement workers or just plain people picked up on the road.

Passing the time in daytime presented no obstacle: one could study mail, discuss the agenda and hold a conference while traveling. Not so at nighttime, after a hard day's labor, on the way back to Tel Aviv or Jerusalem. I would then take the wheel of our black Packard (we had no driver, and would alternate at the wheel) and hurry home.

Eshkol was not addicted to dozing during travel. Those days, too, we had no special reading lamps which do not distract the driver. Reading reports and mail was thus out of the question; and, anyway, I doubt whether we could have summoned up the energy to do so.

Eshkol, therefore, would hum. There was a childhood tune he had picked up in *heder* or from the *hassidim* in his village: a tune which went "bim-bam-bam, bim-bam-bam," pleasant and endlessly monotonous. He would hum for miles, hum and think. If he had a brainwave he would consult me and then go back to his humming.

After half an hour of this, when I felt I was rent through and through by the tune, I would say to him : "For pity's sake, Eshkol, hum a different tune; your 'bim-bam-bam' is putting me to sleep."

"O.K., O.K., forgive me," Eshkol would say, "a different tune it is."

He would search in his mind for a different tune, and start humming something reminiscent of "In this land, delight of our fathers." By the time he had got to "furrow a furrow, sing a song," the tune was already "bim-bam-bam, bim-bam-bam."

"Eshkol, again you're 'bim-bam-bamming.' "

"What, really? Oy vey me," he would exclaim, "I'll hum you a different tune."

He would start on "To the crest of the mountain"; by the time he had reached "climb, climb," it was again "bim-bam-bam."

"Eshkol," I would stop him, "you're turning this tune into just another of your bim-bam-bams."

"Well," Eshkol would groan, "a composer I'll never be. And as for you — you, my friend, are fated to suffer my bim-bam-bams for a long time to come."

In no time at all, I also started humming "bim-bam-bam," and we would hum it as a round, until we got home.

CHAPTER IV

# LOKSHEN AND BLOKONS

If the *ma'abarot* were a quick solution for settling immigrants near the towns and large villages — more complex ideas kept cropping up in regard to agricultural settlement. On the one hand, since the end of the war, we had large arable areas, menacing in their emptiness — whether in Galilee, the Beisan Valley, the northern and southern Sharon Plain, the Jerusalem hill region, the southern coastal plain or the Negev. On the other, the *ma'abarot* and other camps were crowded with multitudes of immigrants, the majority of them without minimal agricultural training. How were the twain to be matched? We knew in advance that the mass of the immigrants would not get used to kibbutz life.

East European Jews were alienated by the "collective" life forced upon them in the *kolkhozes* and *sovkhozes* during their wanderings as refugees in the Soviet Union; people whose homes and families had been broken up and destroyed, who had raised new families and borne children at the wrong age and place, and whose entire thoughts were now given to re-establishing small and private family nests. It was difficult to demand of people like that, that they go to the kibbutz in their thousands.

The kibbutz movements, not without our encouragement, tried their best to attract people. And some they did find, mainly from among the remnants of the pre-war Zionist youth movements, but their number was small.

Then there were the Jews of North Africa, the Yemen, Iraq, Iran and the other countries of the Moslem Orient — families with many children and encumbered with the aged and the elderly. These were patriarchal families, whose world was as far removed from ours and from our concepts of society, as is East from West. We knew that from these, too, we could not fashion kibbutz material.

By way of natural alternative, we turned to the second-largest form of settlement shaped in the pre-state days — the *moshav*. Admittedly, even within the framework of the *moshav* there is a considerable element of cooperation — mutual assistance, centralized purchasing, organized sale of produce — but first and foremost, there is the small house and its piece of land. This was where the Biblical concept could be realized of "every man beneath his vine and his fig tree."

We made up our minds to embark on this road, making it the main highway toward immigrant settlement.

In the beginning, in 1949 and 1950, the work was conducted along two paths — organizing groups of immigrants for the proposed *moshavim*, and planning the form of such *moshavim*.

The *moshav* movement harnessed itself to the job, drawing on its veterans — from among the founders of Nahalal, Kfar Vitkin, Kfar Yehezkel, Be'er Tuvia and other veteran *moshavim* — for information, work and "conscription" inside immigrant camps. Aided by the Jewish Agency personnel, they started forming settlement nuclei for the *moshavim*, i.e. groups of dozens of families who would agree to become farmers.

At the same time, the planners, engineers and architects started working on designing types of houses and farmsteads, and on village planning. These groups of experts were joined by the *moshav* old-timers.

A bitter controversy immediately developed concerning the shape of the village : should the settler's house be attached to the small plot provided for his living, or should there only be a small space near the house for farm buildings, while the plots would be at some distance from the houses. If each settler's plot would be attached to his home, then the distance between the houses would be greater and the village buildings would spread over a large area. On the other hand, it was argued, if the houses were cut off from the fields, with only a half or three-quarters of an acre left next to them — the spaces between the houses would be small and the village more compact.

Until the establishment of the State, the "compact concept" in rural planning held sway, not so much because the planners favored it, but in the main owing to the circumstances relating to land purchases under the British Mandate. It was not easy, and often impossible, to concentrate all the lands next to the homes, while security considerations also dictated concentration of the houses. The old-timers had suffered a lot from the remoteness of their fields and from too close a proximity of their homes; now they gave vent to their repressed desires. They wished to spare the new settlers hardship and insisted that all the fields be near the houses. In this manner, the settler would be able to cultivate his plot without wasting precious time in walking to the fields, and his wife and children would also be able to help him with greater ease.

The architects and planners, on the other hand, advanced the argument that spacing out the houses would distort the shape of the village, extending streets and sidewalks. Women and children would be obliged to cover great distances to get to store and school.

27

As is often the case, the controversy ended with a compromise; but the compromise inclined toward "the spaces." The architects were instructed to plan standard *moshavim* — to accommodate about one hundred families each, with most plots near the homes.

The resultant villages were elongated, their streets extending at times a mile or more in either direction. We soon gave the contours of the new villages such appellations as "towel" or "sheet," but the one which won out in the end was "lokshen," which exactly described the appearance of the long noodle-shaped streets.

The question also arose of the form of the settler's house.

The budget, of course, was negligible. Time pressed, while the engineers came up with plans for houses, for which there was no money and little time to build.

At long last the rural cousin of the shacks and tin and canvas huts emerged — the "blokon." This was a rectangular structure, about six-and-a-half yards long and just under four-and-a-half yards wide built of concrete blocks. It had two rooms and a washing-cum-cooking nook with one faucet; the floor was of concrete; insulating material of sorts made up the ceiling, and the roof was red-tiled. A small wooden outhouse, without drainage, was to serve as the W.C.

All in all, quite an ugly duckling.

The first nuclei were ready to house families at the end of 1949. Tents were put up in the center of the planned village, to serve as provisional accommodation for the settlers, who then constructed the "blocklets" which were to be their permanent living quarters. Each nucleus was duly staffed by an instructor or coordinator from among the *moshav* veterans. These were elderly people, in their fifties and sixties — settlers with dozens of years of back-breaking agricultural toil behind them, years which had left an indelible stamp on body and spirit. They were heavy-footed and stubborn.

However real and keen their desire to help the immigrants, there was not always proper understanding between them and their "pupils." Language pitfalls were not all that stood between them; the misunderstandings derived in the main from the deep abyss between their worlds. They were divided by different concepts, experiences and ambitions.

An occasional source of annoyance for the new immigrants was the tendency of the old-timers to tell stories of the great trials they had undergone when they first came to the country: malaria, hunger, robbers, Bedouin attacks, drought and the like.

The tours conducted for the immigrants to the established *moshavim,* to illustrate the standard of living which could be achieved after dozens of years of toil — these, too, were a psychological error.

The immigrants simply would not believe that the beautiful villas, the perfect farm buildings, the flourishing orchards and plantations had been put up with endless toil where desolation had previously reigned supreme. To them it appeared as though this wealth had been its owners' since the days of creation. Comparison between the old-timers in their flourishing *moshavim* and living conditions in the *ma'abarot* and bare "blokons" achieved an adverse result in the minds of the newcomers. The ideology of the *moshav,* too, was strange to most of the immigrants.

One of the first newly planned *moshavim* was located on the plain facing the (Jordanian) town of Tulkarm, close to the old armistice line with Jordan. Originally called *Bir Burein,* the place was later named Moshav Be'erotayim. The settlers were new arrivals from Czechoslovakia, a few dozen families living in tents. The time was winter, and the rain had converted the heavy soil into mud.

29

We arrived there one evening during those early days, accompanied by some members of the *moshav* movement. The settlers gathered in a large tent to hear what we had to say.

I was seated on a wooden bench among the villagers, wrapped up in their heavy overcoats and wearing mud-spattered rubber boots. Our man stood up and started lecturing in Hebrew on the *moshav* movement and its principles. His language was not understood by the immigrants, and he soon switched to Yiddish; but his Yiddish fared no better. Yiddish was not familiar to these people, at any rate not the speaker's Yiddish. He got carried away by his own eloquence, interspersing his speech with sayings of the founders of the country's labor movement, particularly those of Berl Katznelson. Furthermore, he had brought along a gift for the new settlers — all of Berl's writings in Hebrew, a dozen volumes from which he read out a number of quotations, repeatedly stressing that he was leaving all the volumes as a gift for the newborn *moshav*. The words "Berl Katznelson" and "gift" kept cropping up in his address.

One of the settlers, seated next to me, turned around and asked me in a Czech German-Yiddish: "Excuse me, am I right in assuming that the esteemed lecturer is speaking of the agronomist Katznelson?"

"No," I told him, "he's speaking of Katznelson the ideologist and leader."

"Pity," the fellow said, "I thought he was speaking of the agronomist Katznelson, and that he'd brought us books on how to grow vegetables."

In those early days we came across a hundred and one unexpected problems.

On one of my many trips with Eshkol, we got to Yakhin in the northern Negev. The place had been settled by Yemenite Jews, recently arrived in "Operation Magic Carpet."

We arrived at the *moshav* at the height of the operation of transferring the families from tents to blokons, the construction of which had only just been concluded.

We went into a number of houses, to observe how the people were accommodating themselves. The Yemenite immigrants — small of stature, lean, nimble, shrewd, very alert — soon adjusted themselves to their new situation. With their adaptability, they succeeded faster than other communities in grasping the labyrinthine ways of the "European" bureaucracy imposed on them, and in untangling it by methods all their own.

After our survey, we got down to a general discussion with the settlers, instructors and Jewish Agency Negev staff. The talk abounded with complaints, arguments, requests and demands addressed to us all by the settlers.

Suddenly, one of the settlers stood up, lean, bearded, with long and curly sidelocks, his dark eyes burning as he spoke : he had a big grievance to address to "your honors." "Your honors" — that was us, the whole caboodle of absorption staff and instructors. He had been allocated only one "blocklet," although it was well-known he had brought two wives from the Yemen and had, praise God, four children from the first wife and three from the second. The wives were young and the children small, how could they all live in one "blocklet"? Where was justice? Not only was a grave injustice being done to him, he was also frustrated in his wish to observe the Biblical tenet "to be fruitful and multiply."

One of the instructors pointed out that the "blocklet" was an integral part of the homestead, and that if he were given two houses he would also come into possession of two agricultural units — which would not do.

Glancing about him the Yemenite must have sensed that this time he was faced by more important personages than his instructor, and that this was an opportunity for an on-the-spot decision which would override the local man.

He leveled his glance at Eshkol and said: "Your honor and excellency, it is you people who assert that the new immigrants with their many children are a blessing, and we all pray for more children. Therefore, if you let me have another 'blocklet,' I shall make an important deal with your honors..."

"What deal?" Eshkol asked.

"You will provide me with another 'blocklet,' and I undertake to present the State each year with two children."

Pausing, he then added:

"Two children... at least!"

The upshot was that the Yemenite got his extra "blocklet." It transpired that he could be granted one of the "blocklets" intended for the artisans; these had not been provided with plots of land.

A few months later we received an invitation to attend a circumcision ceremony, with a note appended by the instructor that this was an advance "on account."

CHAPTER V

# THE FIGHT FOR HATZAV

During the first three years of our labors, hundreds of
new settlements were established, mostly immigrant *moshavim*. The "blocklets" covered the face of the land, from the
mountains of Galilee in the north to the southern plains
and the Negev. And across this land, the immigrants conducted their battle of adjustments. In theory, we fought
shoulder to shoulder with them; in fact, they also fought
us. We were the Establishment, to wit, the source of all
their troubles.

Those were the days of rationing and austerity in Israel,
of yoghurt and frozen fish fillet. Rice, meat and eggs were
in short supply; fruit and vegetables — a rarity. Much bitterness obtained in the new villages. Quarrels, fights and
desertions were daily occurrences. In the new *moshavim*
they were at each other's throats, community against community, clan against clan, congregation against congregation;
and all combined against the Jewish Agency personnel and
the instructors.

We put out conflagrations. We would rush from one
*moshav* to the next, with our "fire-fighting" equipment: an
addition to the budget to ameliorate conditions, a few pro-
mises, a lot of forbearance, patience and sympathy. Often we

failed. Villages were abandoned, deserted by their embittered residents, and we brought other immigrant groups in their stead.

On a winter day in 1952, the third year of mass settlement, we were visited in our office by a group from the immigrant *moshav* of Hatzav, situated south of Gedera. They had come to submit a complaint. We were already used to that sort of thing. However, the group of settlers which waited in the corridor until we were ready, made a very special impression on me. Most of them were young men, good-looking, quiet and sad. They neither shouted nor stamped their feet. Their spokesman was a youngster by the name of Ben-Zion Halfon, and he explained to me that Moshav Hatzav had already gone through all the tribulations and sufferings of absorption and settlement. The people of Hatzav no longer shouted or pushed. They realized indeed how busy we were; but they were not going back to the village before unburdening themselves to us.

I had not the heart to send them away. I summoned Levi Argov, my friend and comrade from illegal immigration days; he was then in charge of the Rehovoth district on behalf of the Settlement Department, and Moshav Hatzav came within his authority.

Levi told me briefly of the goings-on at Hatzav; it was a depressing story. We called the Hatzav delegation in and asked to hear what they had to say. Ben-Zion Halfon, aged about twenty, tall and dark-haired, with a pleasant face and dark, lively eyes, spoke for his seven comrades.

There were about seventy families from Libya at Hatzav, who had lived there for close to a year. All the houses were already built. The Jewish Agency had provided them with a small supply of irrigation pipes, as well as cows, mules and implements. They found it still difficult to make a decent living; the land would not respond to their toiling, and they were not accustomed to handling livestock. Yet for all that, there were settlers who were willing to make every effort to master farming and strike roots in the land.

Unfortunately, the village had been taken over by a number of immigrants, whose only wish was to sell the property they had received from the Jewish Agency on the black market, thereafter leaving the village and moving to one of the *ma'abarot* or to an urban suburb.

Ben-Zion explained that the "others" comprised the majority in the *moshav*; they were led by a number of tough and violent men who ruled about fifty families. Whereas the minority was persecuted, beaten and despised for safeguarding the property and equipment entrusted to their care. "It's hell for us," Halfon concluded.

One thing was manifest : if we did not initiate quick action, the village would disintegrate completely. Levi Argov confirmed Halfon's story, adding that instructors from among the old-timers were unable to get a foothold in the village; the toughs had threatened them and finally driven them away. He himself had visited the place dozens of times, trying his utmost to rectify matters, but to no avail. He, too, was of the opinion that the situation was desperate.

We all but made up our minds to propose to the minority that they leave Hatzav and settle elsewhere. We had already experienced such lost cases in the past; we would simply leave the villages to disintegrate, and once all their residents had scattered, we would bring in a new group of settlers.

Suddenly I realized that such a decision would be a terrible insult to the deputation from Hatzav. I also felt it as a burning affront : what was going on here ? These fine youngsters, who wanted to establish themselves in the *moshav* and work its land — they should be expelled, while the group of hooligans was allowed to remain and carry on as though the place was theirs ? Was there no justice ? Any

man shall do what is right in his eyes ? I was filled with anger.
To Eshkol I said :

"Let me have a few days leave. I want to go down there
and attend to the matter personally. Let's not decide anything
until I return."

Well acquainted by then with my stubborn streak, Eshkol
did not argue much :

"O.K., go ahead. See what can be saved from the blaze."

I told Halfon and his comrades that I would be at Hatzav the following day.

I requested them to keep quiet, not telling those of the
"majority" that "Eshkol's assistant" was coming; I would
introduce myself as a new instructor.

I also told them not to count too much on me. I would
see what could be done, and we would then decide. I noticed
a glimmer of hope in their eyes; my mission was to them
as a straw is to a drowning man.

The next day I arrived at the *moshav*. Establishing myself in the empty secretariat hut, I started taking stock of
the situation.

Moshav Hatzav had been located in a prime position, on
the main road from Gedera to the Masmiyeh junction. But
that day it was bleak and sad, like the winter weather en-
folding it. The village was dispersed, somewhat like the letter
U, each arm extending perhaps three-quarters of a mile.
Along its streets sprouted the little "blocklets," bare and
miserable looking. Next to some of the houses the new settlers had put up an assortment of tin huts and clay hovels,
contributing to the dreariness.

Irrigation pipes were scattered near the houses; the fields
lay abandoned and forgotten. There were concrete structures
in the yards, intended for cattle-sheds or stables. Miserable
looking cows and mules occupied some of the sheds; other
structures were empty. I recalled what these cows had looked

like when first introduced from America — veritable "cows of Bashan."

We, the Establishment, not only had we not been prepared for absorption and settlement on such a scale; while we were putting up dozens and hundreds of settlements — we had overlooked the settlers and their requirements. We had sinned against them by a curious blindness and by adopting a routine approach.

For some peculiar reason it had seemed to us that we were dealing with veteran settlers, such as the people of Nahalal or Kfar Yehezkel. We had entrusted them with livestock and equipment, as though they were old and experienced farmers. During those years we had imported hundreds of prime pedigree cattle from the United States, distributing them among the new *moshavim*.

Special emissaries, cattle experts, had been sent out from Israel to the United States to examine the cattle thoroughly — testing them and checking their pedigrees. When they arrived in Israel, the cattle were again examined thoroughly by our veterinary specialists.

Some of the cattle were rejected, and only the very best were sent to the new *moshavim*.

Most of the settlers had never seen such cattle in all their born days. Every cow looked like an elephant to them. The complicated attention this large and gentle animal required confused them and almost drove them out of their wits. Wretched were the cattle and equally wretched the new settlers.

The mules presented an even more serious problem. Our experts had decided that plowing in the immigrant *moshavim* would be done by mules; and if mules, then obviously the best, the largest and the strongest had to be bought. Again the emissaries went out, this time mule experts. In Yugoslavia they discovered a wonderful breed of mule, every last one tall as a cedar and with the strength of a lion. The experts purchased hundreds of these huge mules, brought them over and distributed them among the immigrant *moshavim*.

In most *moshavim*, no land was yet available for plowing and working; but there was the mule in the stable, frightening the new owner. The monstrous animal, with its terrible hooves, required much and expensive fodder. If not properly harnessed, it would kick out in a frenzy. Little wonder that our instructors would find a cow or a mule tethered to the doorway of their hut; the settlers were trying to get rid of the monster that frightened them.

It was not only prime livestock with which the all-abundant Jewish Agency provided the settlers; they were also given mechanized agricultural equipment, and only the very best, of course. Indeed, our norms were crystallized on the scale of the old-established and well-developed *moshavim*; whatever was good for Kfar Vitkin was good for Hatzav. We introduced hundreds of tractors into the *moshavim*: "Fordsons" and "Caterpillars." We purchased combines, "Allis Chalmers" and "John Deeres." We gave no thought to the fact that there was as yet nothing to plow, and certainly nothing to reap. We brought in complex and intricate sets of equipment, calling for expertise and considerable technical skill. The climax was reached with the "Farquhars" — motorized sprayers, the last word in technology; a large yellow tank, riding on a set of wheels and drawn by a tractor, extending out from the body of the tank with as many arms as an octopus, emitting dozens of small jets. The mixing of the spray materials was automatic, and the jet sprinkling was controlled. This expensive apparatus, intended mainly for plantations, was handed out to each immigrant *moshav*. Not one blade of grass had yet sprouted in the fields, not to mention fruit trees, yet there stood the "Farquhar" in the village center. In the best instances, these implements served as toys for the *moshav* children, who used them as swings; but there were other uses as well, as in Hatzav.

All this property, live and inanimate, had been handed over to the *moshavim* without requiring a written receipt. This, too, derived from good intentions. For in the past the settlers, the pioneers of thirty and forty years before, had not been required by the authorities to sign receipts for the property supplied them. It was simply not nice, not the accepted thing. At the present time, too, it did not occur to us to have every settler sign a receipt for property received. We were captives of our good old concepts. At best, we got the *"moshav* committees" to sign; but these committees were absolutely abstract creatures — born one day and gone the next.

Thus I went the rounds of the houses of Hatzav. On every corner I encountered strange and suspicious looks. Towards evening I proceeded to Ben-Zion's house where I was given a hot meal which revived me.

At ten that night Ben-Zion and I left the house. He said to me:

"I assume that tonight, too, the slaughterers will be coming to the rendezvous. I want you to see this with your own eyes."

We advanced slowly toward an abandoned grove on the other side of the road. An hour passed. A truck arrived from the direction of Rehovoth, driving right into the orchard. By the light of the moon, we saw human shadows leaving the *moshav* and setting out in the same direction. I could see that they were dragging a cow with them.

"This is a nightly occurrence," Ben-Zion said. "Black market butchers from Rehovoth, Beit Dagon and Azur come here and buy the cows."

Those were the days of austerity and rationing. The black market in meat flourished.

39

"How many cows have already been sold in this fashion?" I asked in a whisper.

"Twenty at least. At this rate, there won't be a single cow left in the *moshav* in a month's time. Pipes, metal sheets and even wooden boards and tiles are sold by daylight. Tomorrow you can see dealers in building materials coming to the *moshav* and buying up everything that's offered."

Building materials, too, were then strictly rationed. We made our way back to the *moshav*.

I stayed three days in the village, getting to know well its streets and houses. I marked the houses of the "leaders" on a plan of the *moshav* I had brought with me. I got acquainted with the faces of some of them, getting their names from Ben-Zion and his comrades. It appeared that all materials earmarked for sale were collected in the homes of five men, who served as conveyers and intermediaries between the black marketeers and the *moshav*.

Upon conclusion of my study, I proceeded with Levi Argov to the Gedera police station. I told them all about my findings and discoveries. The local police commander was Arie Nir; he did not appear surprised, remarking to me: "I'm aware of the goings-on near Hatzav, and Hatzav is not the only one — it has its counterparts all over the place."

"But I have irrefutable proof, and I want you to go out and arrest the persons selling our property."

"Your property?" Arie Nir exclaimed bitterly. "Where does it say that it's your property? Do you have any document to the effect that the cattle and pipes belong to the Jewish Agency?" Nodding his head, he added, "I have instructions from headquarters not to intervene in these matters, which are developing into a national plague. Failing proof that it's your property, it belongs to the settlers and they can do with it as they see fit."

I then traveled to police headquarters in Tel Aviv. Enter-

ing the office of the Deputy Inspector General, Kutti Keren,
I presented the case to him.

He also nodded his head : "Sorry, we can't take the
chestnuts out of your fire. You committed a stupid blunder in
distributing property worth millions of pounds without
obtaining signatures and securities. First rectify the mistake,
and only then will we be able to intervene within the law."

Returning to Rehovoth, I consulted with Argov. He
asserted that the plague of selling was spreading, and unless
it was stopped, dozens of *moshavim* would disintegrate. "Anyway, what right have we to complain about the new settlers?"
Levi asked. "The biggest black marketeers come from the
old-established settlements, *sabras* through and through. They
are seeking out the new immigrants in their *moshavim* and
tempting them with money and valuables. Who can resist
that?"

"O.K.," I said, "upon my return I'll recommend to Eshkol
that something drastic be done with regard to signing for
property, so that we can get the police into the picture. Meanwhile, though, what's going to happen to Hatzav?

"You know what?" I added shortly. "I suggest that all
remaining livestock and agricultural equipment be forcibly
removed from the *moshav*. Once the *moshav* has been emptied
of equipment, it'll be easier for us to remove those leaders
engaging in black marketeering and their followers."

Astounded, Levi asked: "Who's going to remove it by
force? The good old days are past; this is now a law-abiding
country. What force can you employ ? The police won't intervene."

To which I replied : "We'll try and organize a different
force. We'll get hold of some of the 'boys,' and crack down
on one place. This will quickly lead to rehabilitation there,
and also serve as an example to other places — and, what
is more to the point, will hasten legal and administrative

procedures. Failing a crisis, this affair may drag on endlessly."

We set out vigorously to organize the operation.

Levi conscripted a few young instructors from the district.
I proceeded to the Tel Aviv offices of the Settlement Department, where I held an urgent consultation with Avraham
Ikar, who had at one time been one of the heads of the
*Haganah*, and was now responsible for civilian guard duties
in the settlements, mainly the border settlements. Strong and
sturdy, even at his advanced age, Avraham Ikar could always
count upon some of the tough boys from the outlying quarters of Tel Aviv ever ready for all kinds of "jobs."

After a brief summing up of the situation, we decided
we would bring along about a dozen of the "boys" for the
operation. We also needed trucks and drivers; these we
conscripted from transport companies, where a number of
friends and comrades were employed whom we could let in
on the secret.

On the eve of the action, Thursday, February 21, the
entire "force" gathered in the Settlement Department's offices at Rehovoth. I briefed them on the aim of the operation and the mode of action.

Aim : To remove in one swoop from the homes of the
settlers — all the cows, mules, pipes and agricultural implements, load the entire stock on the trucks, and transport
it to the Jewish Agency's warehouses at Tzrifim.

Forces available : "Property-expropriation" platoon, made
up of Levi's agricultural instructors — eighteen all told. Cover
and protection platoon, made up of Avraham Ikar's "boys"
— fourteen all told. Transport platoon, made up of five
trucks and ten drivers and their assistants.

In command: Avraham Ikar, Levi Argov and myself.

I spread a large map of Hatzav on the wall. The five
houses targetted for action were marked in red, black

arrows marking the roads leading to them.

I instructed the men: "We must exploit the element of surprise and finish the job very fast, before the people recover. We'll set out from here in convoy at 04.30 hours, reaching the approaches to the *moshav* at the first light of dawn."

After the briefing, I went along to Arie Nir at the police station, and told him: "Arie, early tomorrow morning I'm setting out with Avraham Ikar and Levi Argov for a brief excursion into Hatzav."

"Most interesting," he grinned. "Why are you telling me this; surely you don't require a police guide on this excursion?"

"Certainly not; I'm well acquainted with the lanes of Hatzav. I'm also not requesting other help of you. However, I'll be very much obliged if one of your jeeps, with a few policemen, were to patrol the main road from 5.30 a.m."

"And why do you need a police jeep there?" Arie asked.

I said to him: "Arie, we'll be entering into an argument with some of the settlers, a theoretical argument concerning the fate of the property they're holding. As you know, words may lead to words; we're all warm-blooded; suddenly a hand will be raised, and then another, and that comes into your sphere, as far as I understand."

"Oh yes, fights are our daily bread."

"I suggest, therefore, that you be ready, just in case, to separate the combatants tomorrow morning."

Friday, February 22, dawned rainy and misty. Our convoy entered the *moshav* while the people were still wrapped up in their blankets.

The sound of the vehicles awoke the residents. Frightened faces, barely out of slumber, peeped at us from the windows. Our forces split up in no time, according to plan. The instructors and Avraham Ikar's men jumped off the trucks and

43

started collecting and loading equipment and livestock.

Avraham Ikar and I, with some ten others, advanced toward the home of the "big chief," old Ya'akov Reuven.

Ya'akov Reuven's home was some distance from the village center. By the time we reached the house, the old man had managed to summon his ramified family to his yard. When we reached the fence, we saw trouble ahead : Reuven and his three hefty sons stood facing us, screening the wooden gate leading into the yard. Behind them stood about a dozen men forming a "second line," while behind these were some twenty women and children of various ages. The women were screaming and wailing and beating their hands.

Addressing the old man, I said : "We're from the Jewish Agency. We've come to remove from your yard the cows, mules and pipes you received from us, and which you've been selling all along."

Fixing his sharp gaze on me, the old man said not a word. One of the sons answered instead : "You will not enter the yard nor take anything. Everything here is ours."

I said to them: "You'd better let us in. Once we've removed the property, you can pick yourselves somewhere else to live. You have no place in Hatzav."

The son answered: "We'll kill anyone who enters the yard."

As proof of their intention, the sons raised long pitchforks they had removed from the dung heap in the cow-shed. The men behind them were also armed with spades, hoe handles and other implements.

A brief and tense silence descended.

Out of the corner of my eye I observed that at the other marshalling points, the operation was proceeding with relative ease. I asked Avraham to send a runner for reinforcements. I realized straight off that if we did not break the resistance of Ya'akov Reuven and his clan — we would have accomplished nothing. It was also apparent that without a fight we would get nowhere, and that someone had to take the first step.

I waited until the remainder of Avraham's men came running up. I told Avraham : "I'm going to kick the gate open with my foot and try to break in; the minute they raise a pitchfork at me, you break in and lay about you. I'm relying on you and your men for good, dry blows. There's to be no blood spilled here."

"Don't worry," said Avraham, "we are professionals."

I advanced toward the fence. I was on my own. I knew I had no choice. Someone had to start; and, anyway, I was the cause of it all.

I pushed the wooden gate with my foot and faced the old man. The youngest son raised his pitchfork and brought it down forcefully. Instinctively, I put out my hand to grasp the pitchfork which was descending toward my chest. One of the prongs of the pitchfork penetrated deep into my palm, bringing forth a stream of blood.

In a flash, Ikar's "boys" jumped the fence. The sight of the blood aroused them, as it did the instructors who followed them. All of them immediately attacked the men of the Reuven clan.

I felt a sharp pain in my hand. One of the instructors attempted to bandage the hand with the shirt he tore off his back, but the flow of blood would not cease.

Meanwhile, the general melee went on. I saw that our men were doing their job, laying out the Reuven family. An ear-splitting wailing from the women and children accompanied the show.

One of the instructors put me in a car and took me out of the place. On the road we passed the police jeep, which was speeding into the *moshav*. At Gedera I was given first-aid, and from there I was taken to the Hadassah Hospital in Tel Aviv.

The doctor on duty in the casualty ward examined the wound and asked what had happened. Not wanting to go into details, I explained that I worked with immigrant settlements, and during the course of my work had tripped over a pitchfork.

I received an anti-tetanus injection, and a surgeon was summoned. The latter treated me expertly, and after bandaging my hand, advised me to stay in hospital for a few days. By noon of that day I was back at Hatzav.

The police had stopped the fight and a strained quiet prevailed. I met our men in the secretariat hut. Avraham told me that opposition in the yard had ceased within minutes, and all the equipment and cows were removed and then transported to the Settlement Department's warehouses at Tzrifim. The operation had been successful: the number of casualties had been minimal, only one wounded — myself. On the other hand, the shock to the *moshav* people had been complete and profound.

On Saturday, Ben-Zion Halfon came to Rehovoth and told us that there was utter disintegration among the majority. Most were ready to leave the *moshav* and move to urban districts. We consulted with the people of the Absorption Department, requesting that one of their men join us the following day, with a list of urban locations we could propose to those wishing to leave.

On Sunday morning we went out to Hatzav. We settled down in the secretariat, and started calling in all heads of families, according to a prepared list. About thirty-odd families agreed to leave immediately, without argument, and were promised urban housing. Most of them requested to move to Rehovoth.

In the course of the discussion with these people, we realized that they were not all cast from the same die. Some had followed the leaders, and now expressed full regret and a readiness to remain in the *moshav* and follow the rules. Furthermore, some of the leaders, notably the Ya'akov Reuven clan, now appeared to me — after a quiet chat — in quite a different light : persons with initiative, strong, even courageous. The old man, for instance, did not come; he sent his sons. The man was too proud to come and talk to those who, only the day before yesterday, had fallen upon him.

I thought to myself: "What do we want of these people? Is it their fault that there's austerity and rationing in Israel ? Did they bring about the laws of supply and demand of Israel's economy, 1952 ? Did not the black marketeers pounce on them and exploit their situation ? And in the final analysis, it's an open door that invites the thief. What manner of settlement institutions were we ? Where were our brains? Why didn't we get their signatures on contracts?

In short, we agreed that not all families of the defeated "majority" had to leave the *moshav*.

We discussed the matter with Ben-Zion, and I believe he and his comrades also understood and agreed.

Of sixty families, some thirty departed — all those who did not want to or could not stay and live as farmers. Half the families agreed to remain, among them also the Ya'akov Reuven family.

However, we could not dismiss the matter just like that. We got every one of the remaining families to sign a document, composed by us on the spot and copied on a typewriter, using a Jewish Agency form to make it appear impressive. Naturally, it was not a legally binding document.

This is how it went :

Rehovoth District
Settlement Department
*Dear Sirs,*

*In accordance with my pledge to you, given in the course*

*of the clarification between us today, I certify as follows:*

*(a) I undertake to remain at Moshav Hatzav and to be a faithful agricultural settler.*

*(b) I undertake to work my land in an efficient manner and in line with the directions of the instructors of the Jewish Agency's Settlement Department.*

*(c) I undertake to abide by the directions of the elected village council and committee, in every matter relating to their functions in the village.*

*(d) I undertake to pay the dues on time, and to participate in guard duties as per the committee's directions.*

*(e) I undertake to market my produce in accordance with directives issued by the representatives of the Jewish Agency and the Government.*

*(f) I undertake to safeguard the agricultural property, in-animate objects and livestock, and to look after it in accord-ance with the directions of the instructors of the Jewish Agency's Settlement Department.*

*(g) If I do not keep my aforementioned pledges, I under-take to leave the* moshav.

Date . . . . . Signed . . . . . .

Resident of Hatzav

Employing basic Hebrew to explain the contents of the document to the signatories, we secured their signatures to the "official" paper with impressive ceremoniousness.

I returned to Jerusalem a few days later, my hand still bandaged and in an impressive strap. A short while after the Hatzav affair, a special department was set up in the Jewish Agency, known as Contracts and Securities, which commenced to descend on the settlers and sign them up retroactively for every item they had received from the institutions. Gradually, some order was introduced into the mess.

And Hatzav? Did everybody in Hatzav "reach a state

of rest and security"? Not so. Hatzav, like all immigrant *moshavim,* experienced many another upheaval. The throes of acclimatization and growing pains did not pass it by; but in the pangs of growth and development, the young community struck roots, and is now a living and creative fact.

Ben-Zion Halfon is nowadays one of the leaders of Israel's *moshav* movement and Deputy Minister of Agriculture; and who can foretell his future? Ya'akov Reuven saw his farm flourish, and lived to a ripe old age. And Yisrael, his youngest son, the self-same Yisrael who stuck the pitchfork into my hand with such competence, is an outstanding member of the Hapoel football team of Azur.

## CHAPTER VI

# DESCENDANTS OF KING SOLOMON'S MARINERS

The Jews of Cochin hold the belief that they are the descendants of the mariners despatched by King Solomon beyond the seven seas, to search for the treasures of Ophir — which they believe to be India. Those navigators of antiquity are supposed to have established a colony and a trading post in Cochin, marrying the beautiful native girls and setting up their own special Jewish community. Whether there is any truth in the legend or not — it is a fact that a Jewish community of thousands has existed for centuries at the southern extremity of the Indian sub-continent. The Cochin Jews are, apparently, related to the Yemenite Jews. When the Portuguese first set up colonies and trading posts in India some four hundred years ago, the Cochin Jewish community was already an established fact.

Neither in their attractive dark-brown pigmentation nor in language — the Malayalam dialect of the Kerala Province — could the Cochin Jews be distinguished from their native neighbors. As was the custom of Jews in the Diaspora, they, too, engaged in small commerce, brokerage, clerical work and handicrafts. It is one of the wonders of the Jewish people's eternal nature that these few thousand Jews, cut off for generations from the centers of Jewish tradition and creativity, were able to retain their Jewishness within the sea of tribes and nations, religions and faiths encompassing them in this part of India.

They prayed in Hebrew to God, the Most High, Creator of Heaven and Earth. Only a few understood the words of the prayers, but all faithfully kept the basic precepts : circumcision, *bar-mitzvah*, and Jewish weddings and burials. The Indian locale influenced their ceremonials, but the festivals remained Jewish. No great sages or rabbis were produced by the Cochin Jewish community, but their prayer books abounded in moving liturgical hymns, full of yearnings for Zion.

A generation ago, word of Zionism penetrated to the Jews of Cochin. They heard that Jews from all corners of the earth were returning to Jerusalem, rebuilding its walls and restoring its ruins. Zionist emissaries reached Cochin, planting new seeds of hope in the Cochin Jews' messianic yearning. The sons of the Cochin Jews were given magical names : Herzl, Weizmann, Wolfson. The girls were called : Ziona, Tikvah, Geulah.

With the establishment of the State of Israel, the ferment grew among them: the footsteps of the Messiah were indeed heard, and how long were they going to lag behind His footsteps? Jewish Agency emissaries arrived, prompting them to immigrate to Israel. Many sold their movable chattels and sat waiting on their suitcases.

The first of the Cochin Jews arrived in Israel at the beginning of the 'fifties, among them some youngsters who were absorbed in the kibbutzim, within the framework of Youth *Aliyah*. The majority of the Cochin Jews were ready for emigration, and only awaited the signal. All of a sudden trouble descended upon them from a totally unexpected quarter.

It suddenly occurred to Israel's medical institutions that serious diseases, not found in Israel, were prevalent among the Jews of Cochin — filariasis and elephantiasis, diseases common among the South Indian population.

Filariasis, the doctors explained to me, is a disease in which the patient's vascular system is attacked by microscopic worms, which invade the erythrocytes and cause severe and general bodily weakness. Elephantiasis is the infamous elephantine disease affecting the lymph nodes, as a result of which the affected limb — leg, arm or breast — swells to terrifying and distorted proportions.

These two diseases are prevalent as a result of the poor sanitary conditions and undernourishment common among the people of the Far East, and the Cochin Jews were no exception.

Their immigration was delayed, and the matter assumed scandalous proportions. Medical commissions were sent out, bringing back conflicting opinions. Finally, at the beginning of 1954, it was decided by all concerned that the fact that some members of the community were disease-ridden, must not be allowed to prevent the immigration of the entire community. At the same time, it was concluded that, for the good of the Cochin Jews and that of Israel's general population, it would be well for the initial groups of Cochinese to be concentrated in villages, in order to help in the study and treatment of any manifestations of the diseases.

I returned from England that year, replete with studies and thirsting for action. I faced a dilemma : should I continue working at headquarters and in the administration of the settlement activity, as I had done for four years, or should I engage in field work ? I felt that I missed direct contact with the immigrants, with their life and problems. Admittedly, I had often traveled to immigrant centers, spending days and nights among them; but I had the constant feeling that I was as yet unacquainted with my people, and that I had to live the life of the immigrant settlers and their instructors.

That year saw a deterioration in the matter of instructors for immigrant villages. The reservoir of instructors of the older generation was depleted. These men in their sixties, had already exhausted all their strength, energy and patience in hard labor — in their own kibbutz or *moshav*. The three or four extra years they had devoted to instruction in immigrant *moshavim*, among unhappy and refractory people, had been too much for them.

An urgent need thus arose for a new generation of instructors. I did not hesitate much, and upon my return from England I advised Eshkol that I would not continue at that stage working with him or at headquarters, but would seek out an immigrant *moshav* where I could serve as instructor. Going over the list of scheduled immigrants for the following month, I noted that a group of Cochin Jews was due by air from India. They were to be settled in the Negev at Moshav Nevatim.

And that was how I came to the Cochin Jews and to Nevatim.

## CHAPTER VII

# NEVATIM

Moshav Nevatim is located about six miles east of Beersheba. It was first established in 1946 — one of the "famous" eleven villages set up overnight, in the context of the struggle with the British Mandatory authorities over Jewish settlement of the Negev. The place was abandoned by the first pioneers following Israel's War of Independence. In the early 'fifties, a classic immigrant *moshav* was established there, in the "lokshen" style: an elongated rectangle, every one of its four streets as long as eternity. The *moshav* encompassed about eighty "blocklets" with a number of public buildings in its center: a secretariat, a depot, a dairy, armory, cooperative store, synagogue and so forth.

In 1952 a nucleus of settlers was brought to Nevatim, people from Hungary and Romania, most of them ex-concentration camp inmates and other survivors of the Holocaust, who had arrived in Israel along the illegal immigration routes.

The Settlement Department planners' first thought was to base this *moshav* on dairy farming, on the assumption that Beersheba — once it had developed and grown into a town proper — would be Nevatim's milk consumer. It goes without saying that prime cattle were housed in the concrete cowsheds put up next to the "blocklets." It was believed that the settlers of East European extraction would look after their cattle with greater proficiency than their Oriental brethren.

Within a short while, certain unforeseen facts emerged. The soil of Nevatim, which had not been properly surveyed, was unsuitable for the growth of fodder. There was a water shortage because the water came in a small-diameter pipe from Beersheba. The settlers suffered from a feeling of isolation — the Beersheba-Dimona-Sdom road was not yet even a conjecture. Nevatim was connected to Beersheba by a second- or rather a third-class road. There was no electricity and no telephone.

The settlers very soon went over to side businesses : they sold their milk and dairy products privately, rather than through the regular marketing system, and cows went from hand to hand until they landed up with the butchers (those were still years of austerity and rationing).

Moshav Nevatim also became a center for the meat trade — sheep, goats, cattle and other livestock, brought there by the Bedouin. Some of the settlers left, scattering all over Israel; the remainder took over the property of those who had left, becoming big-time dairy operators. Jewish Agency supervision of the half-empty village lessened gradually, until it became practically non-existent.

In 1954 there were about ten families of the Romanian and Hungarian group left in the village, and these were trying to make as much money as possible in the meat business, until they could amass a sufficient amount to establish themselves elsewhere. It was resolved, therefore, to try and advance the departure of the remaining villagers, and to allocate the *moshav* to the Cochin Jews.

I got to know about the happenings in Nevatim at the time in the office of the director of the Negev Region, Munya Kahanovitz. Munya, an expert agronomist who had completed his studies in the United States, was one of the Settlement Department's old-timers.

Sitting opposite me, he surveyed me with growing surprise. "I won't examine your motives, as to why you decided to follow this path," he said, "but if you'll take my advice, before you decide conclusively, grab a jeep and go look over the place. When you get back, come to my room and we'll talk." I took his advice.

The village appeared deserted. The white blocklets stood out starkly against the background of grayish-yellow loess soil, stretching from horizon to horizon. There was hardly a spot of green to be seen in all this expanse, the brown of the thorns holding unlimited sway.

Dust swirled in the empty streets. I walked into deserted houses lacking windows and doors. I noticed that some of them had recently served as sheep and cattle pens; they were strewn with dung, rubbish and bits of straw.

Here and there were occupied houses, and I entered and chatted with the settlers. I told them straight off that it was my intention to serve as instructor to Jews from Cochin, who would be arriving there shortly. I learnt that there would be no difficulty in moving these families, who had been living for months without a school, municipal services, transport or human company.

For my part, I even asked some of the men if they would agree to remain and live with the Cochin immigrants. I considered it appropriate, though, and not necessarily out of sheer innocence, to tell them of the specific state of health of the Cochinese.

The people, the women especially, looked frightened : "Is it contagious ?"

I told them I was no doctor and did not know.

At the extreme north-western end of the village I suddenly noticed a house surrounded by a sea of green — trees and lawns. This was such a complete contrast to the grayishness and forlornness of the bare houses and fields, that my curiosity was aroused. My astonishment grew as I approached the house. This blocklet, with its whitewashed walls and red tiles, looked as though it had been dropped down there from another world and a different vista. Young and luxuriant poplars stood in the yard; there were vegetable and flower beds, and vines intertwined on supports.

There was a half-naked man working in the garden. He was barefooted and bearded, his thick red hair flowing over his forehead and shoulders. He was very bronzed and dripping with perspiration. A veritable Robinson Crusoe. The house door was partially open.

Seeing me approaching, he raised his voice and called out: *"Maria, chiude la porta"* (Maria, shut the door).

Italian in Nevatim — this was something totally unexpected. I saw the woman shooing two youngsters into the house, locking the door behind her.

The thick-bearded fellow, now standing by the fence, beamed at me and said in Italianate Hebrew : "Shalom, my name is Daniel Martinez. What's your name ?"

I shook the layer of dust from the basic Italian I had picked up during my service in Italy in the British Army and in illegal immigration. I answered: *"Bon giorno, Signor Martinez, mi chiamo Eliav; molto piacere"* (Good day, Mr. Martinez, my name is Eliav; very pleased).

His face lit up, and he immediately switched to fluent Italian. Halting the flow of his speech, I asked him to talk at a slower pace. A brief chat elicited the history of his appearance at Nevatim.

He was born of peasant stock in a Sicilian village. Prior to the World War he had been conscripted into the Italian Army, and posted to the Western Desert front. Captured by the British in 1941 before he had seen any fighting, he was sent to a prisoner-of-war camp in Palestine, and very soon became batman of one of the District Officers. He traveled a great deal with his superior around the country, observing how the Jews were striking roots in their land.

At the war's end, after five years in Palestine, he was released and sent back home. Martinez found himself looked upon as a stranger. His brothers, working the poor parental plot, resented his return. With the money he had saved, Martinez was able to marry one of the village girls, but did not succeed in striking roots there. He then recalled the Land of the Jews. Why not join the Jews in one of their new villages, and together with them bring the desolate land to flourish?

Collecting his pregnant wife, his small child and their few possessions, Martinez immigrated to Israel. How did he get into the country? This he refused to tell me, but I presume he managed to infiltrate the large wave of immigration, of hundreds of thousands, to whom Israel had thrown her gates open.

After many adventures and wanderings, he landed in Nevatim. He thought he would be given a plot of land, from which he could extract a livelihood. He certainly never dreamt of a settlement within the framework of the *moshav* movement. Martinez secluded himself in his own yard, like a lone wolf in the forest of the society into which he had landed.

He planted trees around the blocklet — trees which in the opinion of the experts stood no chance of taking root in the loess soil. He used methods employed in Sicily to water his beds, and the vegetables he grew he sold in the Beersheba market. His wife Maria he kept within the confines of the house, as is the custom where he came from, and his three small children — another son had been born here in the meantime — also did not move beyond the fence. Martinez was a Robinson Crusoe on the little island he had built for himself in Nevatim.

He observed the Romanian and Hungarian Jews coming and going. He understood neither why they came nor why they left, but it was not the only aspect of the goings-on around him which he failed to grasp. My telling him that Indian Jews would soon be arriving at Nevatim did not shake him. All he asked was to be allowed to remain and live in peace. I had no need of a course in social psychology to guess that the match between the Catholic Sicilian peasant and the Cochin Jews would not be an ideal one.

Returning to the village center, I enquired about any public property that might have survived. The only thing left was a G.M.C. truck. All the rest had either gone west or returned to the Jewish Agency's warehouses. The truck was in poor condition.

Back in Munya's office in Beersheba, the two of us drew up a new budget for Nevatim. I asked for money for repairing the truck or purchasing a new one; for a budget for initial agricultural projects; for the allocation of work-days in the neighborhood, in agriculture or public works; for new credit for the cooperative store. I described sanitary conditions to him, requesting the immediate installation of new water closets with cesspits in the courtyards. This point bothered me in particular, after having seen the houses there which had been transformed into latrines.

From Beersheba I returned to the Immigration Department in Jerusalem, where I learned that the first large group of Cochin Jews was expected within the week, due in from India by air at Lydda. From there they would be transferred to the Sha'ar Ha'aliya reception camp near Haifa, staying as short a time as possible there. Time was thus pressing, and I had my hands full.

The language question also bothered me. I had it from emissaries who had been in Cochin that English was not a language mastered by the Cochinese. Making inquiries as to whether there were in Israel any "veterans" from Cochin who spoke Hebrew, I immediately discovered that some two or three years earlier a number of youths had arrived from Cochin, and were staying at Youth *Aliyah* institutions or on kibbutzim. I asked to meet them.

Three of them turned up at my office in Tel Aviv : Nissim Eliahu, Nissim Nissim and Avraham Avraham. They were about eighteen, and quite different in appearance and in character.

Nissim Eliahu was short and thin — all skin and bones. He looked as if the least wind would topple him. His skin was a pronounced shade of dark-brown; he had a sharp nose and lively, coal-black eyes, continually darting about like rabbits in their cages.

Also small, Nissim Nissim was a steadier type than Nissim Eliahu. His skin was olive, his face round and handsome, with mild and somewhat melancholy, almond-shaped eyes. His hair was graying, which is common among the Jews of Cochin, bestowing upon Cochinese youth a peculiar appearance.

Avraham Avraham was taller than his friends, broadshouldered, with an athletic and muscular figure. He had Malay features and flowing, very dark smooth hair. There was nothing in his appearance to hint at his being Jewish.

During their two years at kibbutz they had learnt Hebrew. Their spokesman was Nissim Eliahu, and he replied in the affirmative to my question whether they were prepared to assist me at Nevatim, serving as interpreters. They would do everything I required of them for the Cochin Jews and their absorption, he promised. I arranged to meet them at Sha'ar Ha'aliya, the day after the arrival of the Cochin Jews. I felt relieved.

## CHAPTER VIII

# AVRAHAM AVRAHAM TIMES THREE

The day came when I received word that the first forty families of Cochin immigrants were at Sha'ar Ha'aliya. I I drove there immediately.

The Sha'ar Ha'aliya camp was full to overflowing. Thousands of people, a medley of tongues and faces. Men, women and children moving about in between the elongated huts, queuing up for food and in front of office windows, standing in line for equipment and waiting for trucks which were to transport them to different parts of the country. A tumult of dialect and argot filled the air. The patois of the Atlas Jews mixed with Romanian Yiddish; calls in Kurdish were answered in Arabic. Shouts in Hungarian intermingled with conversations in French. And added to all this was the basic Hebrew employed by the immigration and absorption personnel, who attempted to explain, instruct and guide the wanderers through the maze of this modern Jewish Tower of Babel.

I arrived at the hut of "my" Jews, the Cochin Jews.

They were seated on the typical "Jewish Agency beds," bare iron bedsteads, in a posture peculiar to them — the uneasy posture of a people from another world and a different culture.

61

A strange quiet prevailed in the hut; no voice was raised. They were waiting for someone to come and tell them what to do and what was going to happen to them.

I was that someone.

I looked at the small community. There were about forty families.

They were all so thin and — miniature. Most of the men had stringy black beards. They wore faded white cotton clothes, Indian style, topped by Occidental jackets. The women were draped in colorful saris, their black and gray hair worn very long, and carefully and tightly combed on their fine heads.

The boys and girls appeared younger than their age. Like their parents, they were unusually silent; only their black eyes shone in curiosity and wonder.

My young "adjutants" — Nissim Eliahu, Nissim Nissim and Avraham Avraham — who were awaiting me in the hut, greeted me enthusiastically. I passed among the families, introduced myself and shook them all by the hand. I welcomed them and made polite conversation, which was translated by Nissim Eliahu — in a dizzying flow and speed — into Malayalam.

I requested the Jewish Agency clerk to let me have the list of the immigrant families, so that I could get to know them.

Glancing at the list, I was astounded.

The list looked somewhat as follows :


List of immigrant families proceeding to Moshav Nevatim :

| Family serial No. | No. of immigrant card | Surname | Forename | Year of Birth |
|---|---|---|---|---|
| 1 | 70552 | Eliahu | Sara | 1915 |
| | " | Eliahu | Shimon | 1937 |
| | " | Eliahu | Menachem | 1939 |
| | " | Eliahu | Avraham | 1942 |
| | " | Eliahu | Hannah | 1947 |
| | " | Eliahu | Miriam | 1950 |
| | " | Eliahu | Rachel | 1953 |
| | " | Eliahu | Avraham | 1926 |
| 2 | 70553 | Avraham | Sara | 1932 |
| | " | Avraham | Rachel | 1948 |
| | " | Avraham | Rivka | 1950 |
| | " | Avraham | Moshe | 1952 |
| | " | Avraham | David | 1953 |
| | " | Avraham | Eliahu | 1891 |
| 3 | 70328 | Eliahu | Rivka | 1894 |
| | " | Eliahu | Elias | 1934 |
| | " | Eliahu | Solomon | 1937 |
| | " | Eliahu | Simha | 1940 |
| | " | Eliahu | Eliahu | 1912 |

The fourth family was known as the Avraham Avraham family, and numbered six persons. The fifth family was called the Ephraim Moshe family; the sixth, the Eliahu Nehemia family; the seventh, the Nehemia Eliahu family; the eighth, the Eliahu Eliahu family; the ninth, Moshe Ephraim; the tenth, Avraham Moshe; thereafter, again an Avraham Avraham family. An so on and on, almost all possible combinations of the names Eliahu, Avraham, Moshe and Nehemia.

I asked Nissim for the meaning of this. He explained that these were the Hebrew names of the Cochin Jews; they also had foreign surnames, however, which they now wished to forsake and therefore did not record them in the immigrant cards. Asking to hear an Indian surname, I found out that such a name encompassed half the letters of the alphabet, and pronounced by me it sounded as though my mouth was full of pebbles. I preferred Avraham Avraham.

I drew up on-the-spot statistics of the Cochinese families, and fixed the date for their arrival at Nevatim.

Returning to Nevatim, I pressed the Jewish Agency's Technical Department people to install the toilets. I turned over the truck for repair, made arrangements for work for the first days, and marked the empty houses for allocation to the incoming families. I requested and received a number of rifles from the police. Being aware that none of the immigrants had ever handled a rifle, I sought a district commander for the place. The Jewish Agency sent me Ya'akov Levi, a valiant redhead, an all-rounder — driver, mechanic, as well as district commander in time of need. I knew, in fact, that Ya'akov and I would be the only bearers of arms at Nevatim, and that we would have to guard the village to the best of our ability.

The immigrants arrived on the appointed day, and were housed according to my lists. The following morning I summoned all the men for a discussion in the village center, next to the secretariat.

First on the agenda was the distribution of working boots. While still at Sha'ar Ha'aliya, I noticed that the Cochin Jews had arrived in sandals or summer shoes, and I requested that they be measured there for black, high work boots.

A sack full of boots arrived with the immigrants, and I now commenced to distribute them. The first pair I withdrew from the sack had a tag attached, with Avraham Avraham written on it. Three candidates stepped out when I called this name — all Avraham Avraham. There and then I decided that a serial number had to be given to each, and the same procedure was followed in the case of the three Eliahu Eliahus, the three Nehemia Nehemias and all other doubles and triples.

Thus were born at Nevatim Avraham Avraham A, Avraham Avraham B and Avraham Avraham C.

The most important and urgent job, which I wanted the men to undertake immediately, was the cleaning up of the houses and of the village. In this first talk I decided, therefore, to enlarge on the question of hygiene and sanitation.

I started talking, uttering a few sentences and pausing, to allow Nissim Eliahu, my right hand, to translate them into Malayalam. I commenced by saying that "one of the most important things we must all look to in the new village is the cleanliness of the houses, yards and washing installations." Nissim translated my words into the Cochin language; imagine my astonishment as I noticed my audience shaking their heads from side to side, in a negative fashion.

I added a few sentences on the same subject. Nissim translated. The audience continued shaking their heads negatively — this time, seemingly, with even greater enthusiasm.

I whispered to Nissim : "What's the matter, why don't the people agree ?"

"Not agree ?" Nissim wondered. "Of course they agree, and how !"

"If they agree," I asked him, "why do they continually shake their heads negatively, from side to side ?"

Laughing, Nissim explained : "In India, shaking the head from side to side is a mark of agreement, while shaking the head up and down indicates objection."

From then on, I was pleased to see that after every sentence translated, the Cochin Jews moved heads emphatically this way and that, as a sign of accord.

Dividing the forty heads of families into groups, we distributed spades, rakes and shovels, and they scattered among the houses.

Being the only driver in the village, I took the wheel of
the G.M.C., which now became a garbage truck. I moved
from house to house, loading the heaps of rubbish removed
by the men from the houses and yards. When the truck was
full I drove off, a few of the Cochinese hanging onto the
sides. We drove about three miles to the east along dirt
tracks, until we came to a deep ravine into which we dumped
the rubbish.

On this hectic morning, it became evident that there were
quite a few snakes in the houses and yards. Every now and
again I would hear a yell of "Phampha!" "Phampha!" A
number of the Cochinese would immediately jump up with
sticks or spade handles, attacking the snake with professional
skill until they killed it. "Phampha" was the first word I
learnt in Malayalam.

The Jews of Cochin had been expert snake hunters in
their locale. It was a sort of national sport, and here now,
already on the first day, they were practicing it.

Thus the first day passed at Nevatim. In the evening, the
kerosene lamps were lit in the forty houses. The men proceeded to the bare club hut, for evening prayer. Together
with Nissim, Eliahu and Avraham, I visited all the houses to
say good night.

Night descended. Ya'akov, the district commander, I and
my three young aides maintained a guard roster; we watched
in pairs, walking around the houses of the *moshav* armed
with two rifles.

The following days and nights were crammed full. What
did I not do: I was instructor, driver, school teacher; I
attended to employment, food, health. A glimpse at my work
journal of those days might, perhaps, provide an idea of the
problems besetting an instructor during the first days and
weeks of the existence of an immigrant *moshav*.

With the Jewish Agency's Technical Department: Digging the cesspits; color plastering of the houses; screening of windows and doors (at the insistence of the District Health Officer); showers; generator for street lighting; signposts; window panes; trash cans.

With the Settlement Department: Telephone budget, work clothes, advances on account of work, extra work tools, milk allowance for the children, budget for bookkeeper, land survey, budget to buy some mules, petty cash, allowance for ornamental projects and for the Festival of the First Fruits, contract forms for signing by the settlers.

With the offices in Beersheba: Loan of Scrolls of the Law; issue of ration cards; radio set; ritual bath; *Histadrut* membership cards; football and volley ball. Games for school and kindergarten; "Egged" — bus, identity cards; entertainment evening with the *Nahal* troupe from Beersheba; ritual slaughterer for poultry; *mezuzot* for the houses; call by mobile post office; teacher and kindergarten teacher; school furnishing; folk-dance instructor.

With *Hamashbir:* Special allocations of rice and sugar, brooms, kerosene, soap, D.D.T., molders' hammers, iceboxes, fresh poultry and fish, complaint about rotten peas, mule fodder, ice.

With *Kupat Holim* and Ministry of Health: Arrangements for pregnant women, nurse, drugs for clinic, establishment of hygiene and health committee, lecturer on health matters, snake problem, lung X-rays for all residents, *Kupat Holim* health information films.

With Israel Defense Forces: Extra rifles, training allowance, lock and bolts for armory.

The initial period was superseded by a routine. I managed to get hold of two more assistants, Joel and Jonah, both former kibbutz members who had drifted to Beersheba and now agreed to work with me. Joel's main task was the agricultural work roster, while Jonah assisted me in all mechanical jobs : attending to the truck, the water pump, generator and so forth.

Agricultural employment during the first weeks at Nevatim was minimal. We managed to sneak in a few work-days at the Jewish National Fund's Gilat nursery and in various afforestation projects along the Negev roads. The truck would take the men out to work early in the morning, bringing them back at nightfall. Munya provided me with an allowance for ornamental planting outside the houses and along the streets. We brought saplings, the people dug pits and planted trees, and made every effort to tend them.

I bought a few hundred young hens from Kibbutz Hatzerim, distributing them among the families of Nevatim — some fifteen hens per family.

The life term of these Leghorns was very short; neither people nor hens got accustomed to each other. My instructing ability in matters pertaining to the hen roost was quite limited, and I imagine that the translations of Nissim and his colleagues merely added to the confusion.

In short, the chickens very soon found their way to the Cochinese's soup pots. At least, it was a welcome addition to the families' limited diet.

The Cochin Jews' attitude toward me was at first mixed. If I did not know how to take them, they, too, did not understand what made me tick, what my intentions were, and where I was leading them.

In the first days they accepted me as some sort of sahib — the type of British master they had come across in their own country. Nevatim they viewed as a sort of plantation, they being the hired help and I their manager. I could even imagine they would not have been surprised had I walked up to them one day and tested their muscles.

Not many days passed, before they saw in me a most pecu-

liar phenomenon. The sahib loading dung and garbage together with them, transporting the pregnant women to hospital,
living among them and making do with little, bedding down
on a Jewish Agency iron bedstead in the secretariat hut. What
manner of sahib was this ? Perhaps he was not a sahib, but
some kind of fakir?

Gradually, though, the barriers toppled. A contributor to
this was the aunt — Nissim Nissim's aunt, who decided to
"adopt" me.

Nissim's uncle and aunt and their family, the Yosef
Eliahus, were among the first immigrants to arrive at
Nevatim. Yosef Eliahu was short of stature, fairly elderly,
his eyes darting about behind his spectacles, with a sharp
nose — all naïveté and kindliness. But the autocrat of the
household was the aunt, Yosef Eliahu's wife. She was then
in her forties, taller than her husband, with noble features,
her gray hair done up in huge plaits descending over her
finely curved shoulders, her body made as if to be wrapped
in a sari, an olive skin and a large motherly bosom.

The aunt was truly blessed. She had sons and daughters
— big, medium and small. It seemed as if she always had a
babe at her bosom.

One evening, about a fortnight after we had come to
Nevatim, Nissim Nissim came to my hut, followed by the
aunt.

"My aunt has a request," he said shyly.

"By all means," said I, "tell me what it is." I expected
a routine matter, like the hundred and one things which
cropped up all day long.

"My aunt says that she can't bear to see the way you live
and what you eat. Her heart feels for you, and when she
heard from me and from Avraham Avraham, that you have
a wife and baby in Tel Aviv, and that your wife is pregnant,
she knew no rest, finally requesting me to bring you along

to her home and have you join the diners at her table."

All the while Nissim was talking, the aunt stood on the doorstep, her well-spaced white teeth exposed in the wonderful smile she flashed at me.

I tried to get out of it, but without success. We finally compromised, and I agreed to be the aunt's guest for Friday night dinner.

There was a fragrance of incense and Indian perfumes in the blocklet, mixed with the smell of strange dishes and spices. The head of the household and his sons, all dressed in white, and the women in festive and colorful saris, awaited me. After the ceremonial blessing of the wine, in the Cochin Jews' version, we sat down to the meal. Room was made for all of us, and the aunt served her dishes.

From the very first course, my palate was set on fire by the aunt's sharp spices. With the second course and then another dozen or more, my stomach also went up in flames. I was polite and ate all the courses. I tried to quench the fire with the drinks on the table; there were different kinds of brandies, brought by my hosts from Cochin. And very tasty brandies they were, too, but they merely fanned the flames of the curry and other spices.

Somehow or other I managed to get through the meal, and after the Sabbath songs, I thanked the family heartily and set out for my hut. I walked very sedately as far as the fence, but once beyond it, I set off at a gallop for the faucet outside my hut. Turning it on full, I seemed to drink all the water in the Negev pipeline.

After a few Fridays, I became accustomed to the aunt's dishes, eventually even acquiring a taste for them.

The days went by at Nevatim. Special feast days broke the monotony of the workdays : the first circumcision ceremony, at which I was godfather to a baby named Ben-Gurion; the first wedding; presenting the synagogue with Scrolls of the Law; and, at the other end of the line, an evening with the entertainment troupe of a *Nahal* unit stationed in the Negev. On that fantastic evening, it transpired that the Cochin Jews had a store of wonderful songs and liturgical poems, imbued with a yearning for Zion, their melodies blended with Indian rhythms.

The *Nahal* boys taught the Cochinese the song "Anemones bright, tunes of lilies resound; the fragrance of millet is on the Carmel."

And they learned in return: "How beautiful upon the mountains are the feet of the messenger." Truly it was an evening of "the ingathering of exiles."

## CHAPTER IX

# UZZI'S "AUTOCAR"

In the spring of that year, shortly after I had come to Nevatim, Ben-Gurion issued a call to the young people, the second generation members of the veteran *moshavim* and kibbutzim, to volunteer for service in the immigrant *moshavim* throughout the country, with special emphasis on the Negev and Galilee. He called upon them to leave their flourishing farms for a year or two, and to set out with their families to live in the immigrant *moshavim*, to serve as instructors to the immigrants, share their initial trials and help them through the tribulations of absorption.

It was a bold call, addressed to the generation which had fought the War of Independence, people my age, who had already shouldered many tasks: *Haganah, aliyah* and the war. Many of these people had only recently returned home from the Israel Defense Forces, after long years of service.

Some of the outstanding members of that generation in the veteran *moshavim* led the way. And the people answered the call. The second generation of the kibbutzim joined them. The leaders organized the first few dozen volunteers in teams of five, sending them out to the immigrant *moshavim*.

I met them. Many were comrades from service days; I requested that a team be sent to Nevatim as well. They had heard of my pioneer work with the Cochin Jews and were pleased to co-opt me retroactively into their group, and to attach a team of instructors to me.

Thus, on a warm summer day, the gang arrived at Nevatim — Naftali, "Keuke," Ovadia and Orah.

Naftali and "Keuke," both from Kfar Vitkin, were mustachioed, tall, thin, as muscular as basketball players. Orah, too, was from Kfar Vitkin. Ovadia, a member of Moshav Herut, was originally from the United States; crew-cut hair, American style, quiet and industrious. It was immediately apparent that this group would effect changes in the pattern of life and work at Nevatim.

I placed an empty blocklet at their disposal, where they lived in communal fashion, with Orah as the "house mother." The first evening, by lamplight, I told them about my Jews, their ways and customs. I drew on my own experience to explain how to work with and treat these people, who were kindly and disciplined, yet weak and fragile — both physically and emotionally — owing to their being still in strange surroundings.

Together we drew up a plan of work and agricultural instruction in the field. The next day the men set out to work at the head of three groups of settlers. For the first time I felt that the burden on my shoulders was being shared.

It was a pleasure to watch Naftali and "Keuke" instructing their wards in the ABC of agriculture — how to harness a mule, sow, fertilize, irrigate.

The Cochinese also sensed that a new era had dawned at Nevatim. The days of snakes and trash removal were over; organized agricultural work was at hand.

Ovadia undertook the job of firearms instructor. From the Army we received a machinegun and some rifles for instruction purposes, as well as a training allowance. We started putting the settlers through preliminary drill in the maintenance and use of firearms.

And what did Orah do ? Orah was for Nevatim a sort of living example of the end product of girls of the second generation in Israel's established settlements. She was very beautiful, delicate skin, bright eyes, and whenever she laughed — which she did often — she displayed her lovely teeth. Her presence alone was sufficient, but we imposed three more tasks on Orah: to be a mother and housewife to the group of instructors ; to be kindergarten teacher and nursemaid for the immigrants' children; and to act as poultry-farm instructress.

Orah had acquired expertise in poultry farming at her home in Kfar Vitkin. The fresh supply of young hens we brought to Nevatim was distributed by her among selected families. These flocks received superior attention from Orah, who supervised the hen owners, instructing them on the rearing of laying hens.

Come nighttime, having dined at Orah's table — I used to alternate between the aunt's and Orah's cooking — we would fall asleep, exhausted from the day's toil. But not Orah ; she would wait up for her Uzzi.

The entire village, settlers and instructors, was made aware of Uzzi's arrival at Nevatim at the hour of midnight. For with Uzzi came his thirty-ton Autocar.

Himself from a *moshav*, Uzzi was a tall young man, abounding in vitality, youth and muscles. He was employed as a hired driver on one of the giant trucks of the *Mifalei Tovala* haulage company. The first potash output was then being conveyed from the Dead Sea by these trucks. The potash could be transported only by a long and round-about route — from Sdom to Yeruham, then on to Beersheba, and from there north to Haifa. It was a back-breaking trip, lasting two days. And what more natural than Uzzi's overnight stay at Nevatim, where his girl Orah had settled ? If this involved some dozens of miles more for the Autocar and its thirty-ton load of potash — who cared? In any case, haulage of the potash was fully subsidized and in an experimental stage. And if this subsidy also covered a brief encounter for Orah and Uzzi — where was the damage?

Thus, at midnight, we would see the beam of the big truck's headlamps, probing and blazing the way through the swirling dust to Nevatim, the heavy engine roaring and panting from afar. Soon the enormous prehistoric monster would stampede into Nevatim's only street, jarring houses and occupants alike. With a grinding of brakes, the huge Autocar would come to a halt outside the instructors' home, Uzzi jumping out of the cabin right into Orah's waiting arms.

The great desert silence would once more descend upon Nevatim.

Two months later, another group of immigrants from Cochin arrived at Nevatim — some thirty families, recently arrived from India. This time we were well prepared.

A committee from among the Cochin "veterans" was elected to attend to the newcomers' reception. The blocklets allocated to them were renovated and whitewashed. The standard furniture was: iron bedsteads, three blankets per person, a table and chairs, a kerosene burner, a hurricane lantern, pails, pots and a wardrobe. Orah and the kindergarten children decorated every gate and doorway with a big welcome sign, and put flowers in every house.

We scraped together a few pounds for soft drinks, the housewives baked tasty Indian cookies, and the feast was all set.

The reception of the new families was easier, of course. The newcomers were looked after by their relatives, who showed them the ins-and-outs of the village, explaining the nature of the work to them, pointing out the instructors and "Eliahu" (that was me). The Cochinese were simply unable to grasp the name Eliav.

Two women teachers joined our team. After getting an allowance for a part-time bookkeeper, I brought in Alkalay — a clever and sharp Bulgarian immigrant — from Beersheba.

The night watch was now entrusted to the Cochinese, who had meanwhile learned to use firearms. They took the matter very seriously, and it was a veritable danger to enter Nevatim at night. The two watchmen on duty moved as silently as cats, neither smoking nor talking to each other, all in accordance with the regulations we had drilled into them. Anybody entering Nevatim innocently from Beersheba was quite likely to get a heart attack with the watchmen suddenly bursting upon him from behind a house or a fence with their dark faces and burning eyes, fingers ready on the trigger, and gun barrels aimed at the poor visitor's chest. Usually it was a friend of mine or of the instructors, who dropped in for a drink.

Toward the end of summer I felt that the village was well on its way and was in good hands. Tanya, my wife, was about to give birth, and I knew also that new projects were in the offing, projects in which I wanted to share. I decided to bid Nevatim good-bye.

It was a moving farewell party. Everybody was there : the Cochinese in their festive apparel, the instructors, some of the Settlement Department people from Beersheba accompanied by Munya Kahanovitz, the storekeeper, the bookkeeper, and the driver. There were many nice things said about me in Malayalam, and I gave back measure for measure, with Nissim Eliahu and Nissim Nissim translating the speeches, and well-wishes flashed back and forth.

At evening's end, one of the elders rose and presented me with a golden ring bearing the State emblem done in blue cobalt, saying: "Dear Eliahu, two such golden rings did we make in Cochin, prior to our departure for Israel. One we have set aside for the President of the State of Israel, and we are holding it for the time we go on a pilgrimage to Jerusalem, when we shall present it to him in person. The other ring, we said to ourselves, we will give to the person who helps us and deals kindly with us in the new land. You are that person."

I shook the old man's hand and put the ring on my finger.

The instructors gave me a farewell present of their own devising — a folded scroll containing the text of a "telegram." Their message read as follows:

Serial No. 1 — Date 8.26.54 — No. of words 51 —
Post Office Nevatim:

Dear Eliav,

We wish to express to you all our heartfelt thanks, on your leaving us. From the day you came to us, at Sha'ar Ha'aliya and until this very day, you have been devoted to us. Parting from you is difficult after all these months. And we hope you will at least visit us at stated intervals and will remember us wherever you may be; and we shall not forget you, and shall try to follow the path you have set out for us.

Wishing you success in all your endeavors,

Residents of Nevatim

## CHAPTER X

# LAKHISH

At the end of 1954, David Ben-Gurion and Levi Eshkol were still on good terms.

Ben-Gurion was then at Sde Boker, having resigned from the Premiership a short while before. But he still toured new regions and immigrant centers; and his desire for first-hand knowledge of settlement problems did not cease. Levi Eshkol was then Minister of Finance in Moshe Sharett's Cabinet, and was also head of the Jewish Agency Settlement Department.

Eshkol summoned me to Jerusalem: "B.-G. has just spoken to me about settling the region extending from the Gaza Strip to the Hebron hills. He has just returned from a tour of the area, full of enthusiasm. Go and visit him at Sde Boker and listen to what he has to say."

After arranging an appointment with the Old Man, I took a jeep and descended from Jerusalem to Sde Boker.

Paula, B.-G.'s wife, was standing at the doorway of the green hut. She interrogated me as to my purpose in coming and warned me not to stay too long. Only then did she allow me into Ben-Gurion's room.

Ben-Gurion was seated at a table full of books, manuscripts and papers. He already knew about my doings at Nevatim and received me cordially. First, he questioned me on my experiences among immigrants. He then proceeded to describe his impressions of the vast unpopulated areas that he had seen in his travels throughout the South and the Negev. He laid particular stress on the need for settling the empty lands between the Jerusalem highway and the Negev, between the Gaza Strip border and the Hebron hills.

I told him that I had spoken with Eshkol, and that a decision was pending on setting up a special team for developing this region. I was the candidate for project coordinator, I informed him.

Ben-Gurion seemed very pleased. He already had in front of him the person entrusted with planning and implementing his vision of settling the new region.

Within a matter of days, the Settlement Department's administration set up a new settlement region, Region "E" (four settlement regions then already existed : North, Center, Jerusalem and the Negev), and I was to be its director.

I immediately commenced the search for a name and a team. I contacted the office of the President, Yitzhak Ben-Zvi.

Yitzhak Ben-Zvi, amateur geographer and historian of the Land of Israel, was extremely pleased at being consulted on such a question. I showed him the region's boundaries on various maps. Ben-Zvi then took out his own historical and archaeological maps, looked into the Bible and the Concordances. He seemed lost in thought.

At long last he raised his eyes to me, and through his spectacles I glimpsed the glint of discovery.

"The region you are about to settle was part Philista and part Judea. In the center of this region, where the coastal plain meets the slopes of the hills, Jews fought Philistines for centuries.

"The large Jewish fort in this region was the town of Lakhish. Excavations conducted there, not so many years ago, provide conclusive evidence of the importance the Kings of Judah attached to this fortress town.

"I propose you name the entire region Lakhish, thus reviving the memory of one of the fortress towns of Judea."

There was something tough, fundamental and rocklike in the name. It evoked ancient scenes of Jews battling Assyrians, of beacons and banners, shouts and battle cries.

I thanked Ben-Zvi warmly for his idea, and left with the name The Lakhish Region.

Now I was faced with the problem of finding my team and my work site.

Past experience had taught me the importance of assembling a team from the very first day, of all the experts in the various planning branches. It was essential for such a team to be together under one roof. I was aware that lack of coordination between the departments and the planners had caused us great trouble in the past. Often the agricultural planner was not aware of exactly what the hydrologist was doing; the latter did not know the plans of the civil engineer who did not grasp the ideas of the highway engineer; and so on and so forth. This situation had been the cause of unnecessary mishaps and much annoyance, and, what was worse, had marred the plan from the outset.

Now that I was given the authority to plan a new region from scratch, I resolved to avoid duplication by concentrating all Lakhish planning and operational personnel in one office. This would neither be located in Jerusalem nor Tel Aviv, and not even in Beersheba, but would be in a place nearest the project. The choice immediately fell upon Migdal-Ashkelon, which already numbered a few thousand inhabitants.

At first sight, my plan appeared quite revolutionary and rather extreme: to compel engineers, architects, agronomists and their like, not only to plan a new region but also to act as a sort of vanguard. But would we find people prepared to accept this condition as well as to move to a new development town?

This idea, though, stuck fast in my mind, guiding me in my search for personnel.

I commenced by looking for two assistants : one to be in overall charge of planning, the other to be chief supervisor of operations. The choice was not difficult. I was well acquainted with the senior staff of the Settlement Department ; most were veteran comrades. I summoned two of my friends, who were among the best employees of the Department, to a meeting at Café Hermon, opposite the Jewish Agency compound in Jerusalem.

Levi Argov and Benny Kaplan arrived on time. They had already sensed what was in the wind.

## CHAPTER XI

# THE TEAM

I knew Levi Argov in the days of illegal immigration. When we first met on the escape routes out of post-war Europe of the survivors of the Holocaust to Palestine, Levi was then in his late twenties. Strongly built, he had a round face, with raven-black unruly hair descending over brow and eyes.

Born in Czechoslovakia, Levi immigrated to Palestine before the Second World War with a group of the *Hashomer Hatzair* youth movement. Before long, he joined the *Haganah* and the *Palmah*. During the war, he was one of the handful of the wonderful boys and girls assigned to parachute into occupied Europe for special tasks.

Immediately after the war he started organizing the "escape" routes out of Eastern and Central Europe, his bases of operations located mainly in Hungary and Czechoslovakia. His non-Jewish, almost Slav appearance provided him with good cover in this underground work. It was in these circumstances that we first met.

During the War of Independence, Levi — who had made excellent connections with the Czech authorities — was entrusted with organizing the airlift to Israel of the arms we had purchased from Czechoslovakia. Immediately after the war we met again in Israel. I was already working with Eshkol. I encountered Levi in a small café in Tel Aviv; he looked sad.

"How are things, Levi?"

He told me of two events that had taken place in the meantime: he had married Orah, his sweetheart, and he was about to leave his kibbutz and was seeking a new way of life.

"Come and work with us, Levi," I said. "We're starting a big thing. Masses of Jews are arriving in Israel, and we need people like you. We're settling them on the land. With your background, in the agricultural and organizational sphere, you'll find your place with us."

And that is how it began. Levi worked for four years on settling people in the Central region and in the South. Rehovoth was his headquarters, where he did a first-rate job.

I now called upon him to be my assistant in Lakhish, and he joined me without hesitation.

My second assistant was Benny Kaplan, who had only recently been discharged from the armed forces. Prior to that, he had completed his agricultural studies, and already enjoyed a reputation for his work in agricultural planning.

Benny, who was over six feet tall, had wise eyes and a high forehead plus a spreading baldness, which gave him a "learned" appearance. As indeed he was, with all his knowledge of agronomy and agricultural planning. Benny combined the cleverness and erudition of generations of wise and intelligent East European Jews with the kindly cynicism of the *sabra*.

Levi became my assistant for operations, while Benny looked after the complicated area of agricultural planning: he, in fact, held the key to moulding the shape of the future region.

The *Afridar* quarter was then in the process of being built in Ashkelon, and a handsome office building had already been put up in its center. We approached Dr. Sonnabend, Director of *Afridar*, and succeeded in renting a few rooms in this structure. We decided to move our families to Ashkelon as well so as not to waste time commuting.

A small two-room apartment fell to my lot at No. 11 Rehov Havradim. We packed the few possessions in our roof apartment in Tel Aviv, and taking the two children — six-year-old Zvika and two-month-old Ophra — we proceeded south to Ashkelon in my new jeep station wagon.

Benny set to work like mad. He had a theory that it was imperative for members of the founding team to have on hand as soon as possible a primary, inclusive blueprint for the Region.

"If we possess the first outline," he said to me, "we shall be kings here. Whoever holds the first planning chart, is the one who sets the pattern. All arguments, arrangements and compromises have to be made with him." In order to crystallize such a blueprint, Benny pushed his men at a terrific pace.

We exerted pressure on the Settlement Department and other institutions to place at our disposal the best people necessary for planning and operations. We sensed this would be our test : it was imperative for us to assemble the best engineers, planners and organizers. We had to bring all the staff to Ashkelon and place them under one roof — let them work together, live together, "feel" the mission and the ground together; then — we felt intensely — they would also reach good and agreed solutions, and would not get in one another's way. Past experience had taught us that dispersal of the planning invited disorder and confusion, which later harmed settlement and settler alike.

Classic cases of disorder came to mind. There was the known case of the *moshav*, with all its houses ready to move into, when it transpired that — according to the national hydrologists — half of the village houses stood on the site of a proposed artificial reservoir.

I recalled a trip with Eshkol to immigrant settlements en route to the Negev. We noticed a group of laborers laying the foundations for new houses. Eshkol exclaimed: "But they're building in the bed of the *wadi*!"

Approaching the foreman, a real veteran, we told him: "*Haver*, you're building in the *wadi* bed; the first flood will sweep away the houses, occupants and all."

The foreman, who knew Eshkol, answered: "What do I know? I contacted the engineer in Tel Aviv and warned him, but all he said was: you work according to my plan and don't ask too many questions."

Eshkol was furious. Without more ado, we got the foreman into the car and brought him to the engineer in Tel Aviv. Eshkol raised such a scandal there, until the very doorposts quaked.

The memory of these and other disasters faced us when we approached the task of setting up our team. Every new man was told: "Dear friend, if you want to share in the task, come and work with us for a year or two in Ashkelon. Moreover, bring your family — we'll be one family and do the job as one."

Thus we assembled the rest of our senior staff.

This was how Hisdai Haviv, the Region's treasurer, came to us. I knew that the big project would call for a first-rate man of finance, who would provide budgets and checks, and would see to it that our machinery was properly oiled and did not throw out balast for lack of money. I wanted a person on whom I could rely one hundred percent, who would relieve me of the entire financial burden. Haviv was then working at the Jewish Agency's Finance Department in Jerusalem. Turkish by birth, he was of medium height, sharp of movement and thought, very alert and with a quick grasp of things. After two or three meetings he joined us. He was a veteran of the Israel Defense Forces. He now moved to Ashkelon with his family, and started setting up our financial wing.

I told him immediately: "Haviv, experience has taught me that our Achilles' heel is the orderly payment of the settlers' wages. You will never have sufficient funds for everybody at the right time, but woe betide us if the wages are not paid promptly, for that affects their very food."

I told him what went on at Nevatim when I was instructor to the Cochin Jews. The same scene was repeated every fortnight: I would submit wage bills to the Negev Region's accounts department at Beersheba, only to be informed that there were no funds to defray the bills, or that there was only half or a quarter of the required amount. Not only did we contravene the major legal prohibition against holding back wages; our people were actually on the verge of hunger, and could not afford to buy a loaf of bread at the cooperative store. The same situation obtained in many other *moshavim*.

Since I was considered a privileged instructor ("Eshkol's former secretary"), and as I knew personally all the Region's directors, I would dash like mad to Beersheba and ask for charity. I would barge in on the bookkeeper, go through the papers, and practically force him to give me a check to cover the laborers' wages.

"This will not happen in Lakhish," I told Haviv.

We evolved an idea that justified itself in the course of time. We printed large red labels, and on them, in big letters, was the legend: "URGENT! LABORERS' WAGES! DO NOT DELAY!" These labels were given to all instructors and foremen in every village. When submitting accounts, they would attach this label. From then on, Haviv instituted a system whereby these red accounts received top priority over all others, and there were always funds to meet them.

A number of people joined us. Rafi Gurevitch, a leading hydrologist, who, having completed his studies in the United States, had gone to work for the Israel Water Planning Authority. He was gregarious and had a pronounced sense of humor.

Shmuel Urbach, an engineer from the Jewish Agency's Technical Department, who had already to his credit long experience in civil engineering; and Shmuel Gerstenfeld, a young and capable architect, who was appointed regional architect and moved with his family to Ashkelon.

Pinhas Sussman, a dauntless *Palmah* fighter, who had only recently completed his studies *cum laude* at the Rehovoth Faculty of Agriculture, who joined the team as Benny Kaplan's assistant.

Arie Meir, born in Germany, a young and valiant kibbutz trainee; behind his orderliness there was a kindly and loving person.

Gershon Fradkin, from Moshav Yarkona, with his rich background in security work. He was put in charge of instruction.

There was Munya, short and agile, one of the *Haganah*'s first wireless and communications men, who took his place in the Region's quartermaster section.

And then there was Shella, the Region's secretary. I first met Shella Peled in the *Haganah*'s Intelligence Section. Volunteering for work with the Lakhish team, she also moved to Ashkelon with her family.

Shella, with her open face and laughing eyes, lost no time in taking charge of us, of our journals and timetables. The Region's staff was quick to toe her line.

There was feverish activity from the very first day in Ashkelon. Concentrating the team in one spot left its mark immediately on the quality and tension of our activities.

Every morning, we were already deep in work at a very early hour. The surveyors were out in the area; agronomists, hydrologists, civil engineers, architects, soil surveyors, electricity engineers, highway and telephone engineers, quite a number of other specialists and professionals, and all engaged in the settlement and population of the land. They descended like locusts on the myriad desolate acres of the plains and hills of Lakhish.

Jeeps, command cars, trucks and station wagons set out at dawn for the territory, returning at nightfall. We occupied some twenty rooms in the center of Ashkelon's *Afridar* quarter, and next to them we also put up four prefabricated huts. These were the Region's offices, and they were soon to be a veritable beehive.

It was imperative to introduce some order into this confusion. I organized a weekly and then a daily work roster. A giant timetable was hung in my room, and on it were drawn in color the networks of our widespread activity.

The engineers' and planners' offices were covered with charts, sketches and aerial photos. The men burnt the midnight oil, after hard labor in the field, putting the results of their work on paper and on the charts. We instituted weekly conferences at which everyone reported on what had been accomplished during the past week in his sphere of responsibility. Activity was coordinated all the time. Weekly reports were distributed among the Region's personnel, while monthly reports flowed to the Department's management in Jerusalem.

In all this hubbub we did our best to maintain work and office cleanliness; and I do not mean just physical cleanliness, although I was very strict about that as well. I would go through the staff rooms almost every evening, together with Shella, checking the state of cleanliness and whether the charts and papers were sorted out and in order. We introduced new and functional furniture, and provided the staff with modern and easy-to-handle instruments. Our furniture was simple — work desks, chairs, shelves and boards for the charts. I did not allow a single armchair in the offices, lest the occupants sink into them and be reluctant to get up again.

I hated pictures on the walls, especially the *Keren Haye-sod* type — "Deganya's early days," "Nahalal's early days," "*halutzim* dancing the hora," and the like.

I issued instructions that there were to be no desks with drawers in the offices, since it is the nature of drawers that they turn into rubbish dumps, with a collection of papers receiving no attention. I insisted that there be no accumulation of unanswered letters.

No tea and cakes were served every half hour at our conferences, as is customary, only cold drinks, and even these were rationed. I hated long conferences, with torn notes lying about on the table and overflowing ashtrays. Our desks had to be clean.

The prolonged talkative conferences, during my work with Eshkol, had implanted in me a loathing for eating at meetings.

This was how those conferences would proceed. The subject for the agenda would be, for instance, "Undernourishment among children in the *ma'abarot.*" Thirty persons would assemble, every one of them with something to say on the matter. Barely quarter of an hour would go by, when steaming tea and cakes were served. All talking, all taking sugar, all stirring the tea and filling the room with the clinking of spoons.

Bereft of tea, all glasses would then be filled with scraps of paper and remnants of cake. The table would be littered with crumbs. Discussion of undernourishment would proceed.

Within half an hour, in would come the servers again

with bowls full of sandwiches.

Ever since, I developed a repugnance for such conferences, and I swore to myself: if ever I am in charge of any project, I will not permit such habits. Indeed, no such conferences were ever held in Lakhish.

Throughout those fall and winter months we lived the planning of Lakhish day and night, in the field and in the office, during our joint meals, and even with our families. We grew closer to each other and formed a small and compact society, whose life's purpose was — Lakhish.

## CHAPTER XII

# NAMES COMMITTEE

A custom prevailed among the Settlement Department planners, in time becoming a fixed tradition: when marking new settlement points on the map, they would give them "provisional names."

The choosing of permanent names was a long and nerve-wracking process, undertaken by the Names Committee affiliated with the Prime Minister's Office — a most distinguished, a most learned, and a most sedate committee. Any name coming forth from this committee was certain to have gone through months of gestation and molding.

That was why the planners were obliged to improvise with provisional names, usually basing themselves on the name of a nearby Arab site or ancient Jewish site. And if they required several new names for the same neighborhood, they solved the problem in a simple arithmetic fashion : *Taanach 1, Taanach 2, Taanach 3, Taanach 4,* ad infinitum.

The settlement authorities cannot dispose of as much time as can the Names Committee. Meanwhile, the new settlers have arrived and have settled at *Jullis 5* or *Taanach 7,* sensing well that the name of the village is only provisional. Taken in conjunction with the other transient manifestations involved in mass new settlement, the impression of impermanence grows — everything is transient and fleeting : the

new immigrant settler and his new occupation, as well as
the village which has no real name, being indicated by a
number.

We decided that in Lakhish it would be otherwise. Here
we would provide permanent names for the settlements, even
prior to the settlers' arrival.

With a map of Lakhish in my possession, on which were
marked some twenty settlement points to be established in
the Lakhish Region, I proceeded to the government Names
Committee in Jerusalem.

The Committee had convened especially to consider my
request. It was no light matter, assembling about a dozen
octogenarians and men of the age of counsel, each important
and outstanding in his own sphere : geography of the country, knowledge of the Bible, mastery of the language, and
so forth.

I presented the map with the proposed new Lakhish settlements to the Committee. They rained innumerable questions on me : where will there be a *moshav* and where a
kibbutz, where will immigrants settle and where natives, what
type of farm will there be ? And other scholarly questions.

For my part, I requested — practically begged — them
to decide immediately on compact Hebrew names for the
future settlements. There were two types of names I dreaded
— the *schnor* type, viz. to perpetuate donors, and names
from the *Haftarah*.

On the one hand, the Names Committee was pressed by
Jewish and Zionist institutions — *Keren Kayemeth*, *Keren
Hayesod*, and other fundraising bodies — to perpetuate
their donors and communal workers. On the other hand,
those Committee members erudite in the Bible and *Mishnah*
tended to conjure up Biblical and *Mishnaic* names for new
settlements which chanced to be established in the vicinity
of settlements and sites of antiquity.

## Names Committee

The first name I asked them for a decision on was for the town due to rise in the center of the Region. And since this town was to be established near the *tel* of ancient Gat (actually, following the archaeological excavations conducted there in the course of time, there have been increasing doubts as to whether this was the actual site of Philistine Gat), and whereas the name of the *tel* had already been bestowed on Kibbutz Gat, the members of the Committee decided — after much careful consideration and deliberation — that the town be named Gitah.

"Why?" I asked. "The place will be a laughing-stock. Gitah will be called 'Gitta.' The inhabitants will say: I live in Gitta, sister of Ritta, or of Ditta... Gentlemen, have pity on the town, which will eventually assume fair proportions, and give it a decent name. Why not call it, simply, Gat?"

"But there is already a Kibbutz Gat," they objected.

"Fine," I said, "there'll be Kibbutz Gat and Ir (town) Gat."

"Ir-Gat doesn't ring well," said they. "But why not Kiryat-Gat?"

"By all means," I answered, "Kiryat-Gat is a fine and compact name."

That was how Kiryat-Gat came into being.

While the going was good, I gave the Committee no respite until I had extracted from them, in one go, a parcel of wonderful names for the Region's settlements :

Otzem, Shahar, Nogah, Zohar, Nehorah, Sde David, Nir-Hen, No'am, Eytan, Sha'anan, Or Ha'ner, Sde-Moshe, Lakhish, Amaziah, Nehoshah, Geffen, Tirosh, Beit-Nir, Sdot-Micha and Luzit.

## CHAPTER XIII

# LOCATIONS COMMITTEE

Question : How does one set about determining the location of a new settlement in Israel ?

Once Benny Kaplan and his men had combed the length and breadth of the Lakhish territories, measuring the land virtually step by step, they proceeded to draw up a comprehensive chart of squares, every square containing a village. The square determines the number of farm units the village will contain unrelated to the type of village — kibbutz, *moshav*, etc. Moreover, the regional planning team does not have a free hand in locating the village within the agricultural square; this involved quite a procedure.

There are other "landlords," claimants and counsellors, representatives of institutions, who have a say in the matter. In short, location of a new settlement is determined by the special committee. This is how it had always been, and for all I know this practice may go on forever, and we had to bow to it in Lakhish as well. One thing was for sure, though, and this too I learned from experience: it is simply impossible to bring together ten experts, each with vested interests in his own sphere, and hope for them to reach agreement in the first instance.

The locations committee is composed of the following :

1. The Director of the region.
2. The Assistant Director of the region (agricultural planning) and his staff.
3. The Regional Architect and his staff, and a representative of the Technical Department at Head Office, Jerusalem.
4. The Chief Hydrologist and his staff.
5. The representative of the Interior Ministry's Planning Department and his staff.
6. The representative of the Israel Defense Forces and his staff.
7. The Road Engineers, on behalf of the Public Works Department.
8. The Electric Company's representative.
9. The Government Antiquities Department's representative.
10. The representatives of the various settlement movements, acting for the future settlers.

A dozen vehicles — jeeps and command cars — accommodate the whole caboodle, equipped with maps, charts, sketches, binoculars, surveying instruments and, of course, sandwiches and drinks, and the convoy sets off for the specific square wherein the village is to be sited.

We descend from the vehicles. Benny lectures the entire audience on the area and its agricultural possibilities. A hundred and one questions are fired at him, and he does his best to answer. We are standing on a hill in the heart of the territory, and Benny points to two further hills, one to the east and one to the north.

Benny says: "On one of these three hills — indicated on the map as 'A' 'B' and 'C' — it will be possible, in my opinion, to establish the village. We have to decide which."

We start reconnoitering from one hill to the next. First we all proceed together, in one large group. Within minutes, though, the pack breaks up into small groups of experts, with their maps and charts. The hydrologists examine the possibilities of laying water pipes to the settlement, the architects inspect each hill with an eye to building, the road engineers discuss access roads to the village and their junction with highways.

The archaeologist is an exception. This fellow brought along a load of ancient tomes and time-worn charts; he also brought with him a small spade, a hoe and cloth bags. He is the most dangerous of the experts. I know this from experience, and I accompany him, chatting with him in an effort to distract his mind somewhat from the secrets of his terrible science.

We climb up the base of hill 'B.' The archaeologist's eye flits over the surface, he picks up shards which never saw better days, marks them and places them reverently in the appropriate bag. "These are shards from the Ancient Bronze Age," he explains benevolently, "what we refer to for short as A-B." Farther up the hill he digs with his little hoe and discovers some more shards. "These, apparently, date to the period of the Kingdom of Judah," he says with undisguised pleasure. "The matter calls for further investigation."

The greater the accumulation of shards in the archaeologist's bag, the more my apprehension grows. And when, bending suddenly, he raises aloft a number of ancient coins, which he announces are Byzantine, I feel that the settlement prospects of hill 'B' are declining fast. And who knows what he will find at the top!

My apprehensions are realized. Up top, with his little implements, he uncovers some foundations of Roman structures — "but this, too, requires careful scrutiny and investigation." Addressing me, he says gently, "You realize, of course, that at this stage I cannot approve building on this hill. We must excavate in order to investigate the structures; we may find an extremely interesting dig here." I recognized the theme; a tough argument was ahead of me.

Nor does hill 'C' disappoint the archaeologist. There, too, he finds shards, coins and the remains of fascinating structures.

I turn on him angrily: "Who invented you archaeologists anyway? If we're to proceed in accordance with your findings, then we can't settle the country at all. Is there a hill in this country without shards and coins? After all, who has not inhabited this land? Stone Age flintworkers; Ancient and Middle Bronze Age man; Canaanites, Philistines and Israelites in the Iron Age. The country was invaded by Egyptians, Assyrians, Babylonians and Persians; Jews resettled here in the Second Temple era; again the country was invaded, by Greeks, Romans and Byzantines. Arab, Crusader, Mameluke and Turk, in turn, conquered the country and settled it. Finally the country was conquered by the British, and Jews came and settled here again. And where, do you think, did they all settle, if not on those hills? The same logic directed them to hills 'A' 'B' and 'C.' And where shall we now put up the houses? In the valley? On the plain? Let them be swept away by freshets in the *wadis*? Let the people suffer the warm and humid air down below, rather than benefit from the fresh air and breeze up above? And what of their security?"

Once again we assemble on hill 'A.' After a reviving drink and sandwiches for refreshment, the discussion commences.

Architect: "I fancy hill 'B.' It's a very nice hill; it has a long spine, extending about 1,100 yards, with two lovely, shorter projections. I believe it could easily accommodate about one hundred houses. The village will be elliptical in shape, the communal buildings and utilities in the center. Hill 'A,' on which we are now standing, does not at all appeal to me, since it is some two miles from the proposed rural center. The children won't be able to walk to school, and there's not much point in a rural center too far removed from the settlements it is designed to serve. Hill 'C' is too small to hold the requisite hundred houses."

Hydrologist : "There's no doubt in my mind that the only hill suitable for settlement is hill 'A,' the one we're on now; only to this hill will we be able to supply water easily and cheaply from the main line, which is planned to pass about 550 yards from here. Any other alternative would involve a much larger outlay; we'd have to lay another 1.5 to 2 miles of pipes, and where's the money to come from for that?"

Road Engineer: "Gentlemen, what are you going on about? The only obvious site for the settlement is hill 'C,' and none other. Only hill 'C' can easily be joined to the planned regional road, and from it to the State highway. Should we, Heaven forbid, put up the village on either hill 'A' or hill 'B,' the access road, culverts included, will cost about a quarter of a million pounds extra, and I have no extra budget on which to draw."

Now comes the turn of the Israel Defense Forces, represented by a lieutenant-colonel and a major. Lt.-Col. Harsina is a *Haganah* veteran who has already been through every single settlement in the country. The major is known by the nickname of "Ephraim the goat," due to his wonderful ability to jump and leap from hill to hill ahead of everybody else.

Say' Harsina and Ephraim the goat: "Both hills 'A' and 'C' are out, absolutely disqualified, security-wise. 'A' is located too far away from a possible invasion axis; the road cannot be blocked from it with flat-trajectory weapons. 'C' is not high enough; an assault force could easily scramble up its sides and capture it. Hill 'B,' on the other hand, would appear to meet requirements; it is located close to the road and the axis of movement."

The architect's face lights up, but his satisfaction is short-lived.

"Yes," says the Army, "hill 'B' is good, but on one condition: we don't agree that, in addition to the central ellipse, the architect should build on the two small projections. There the houses would be isolated, an excellent target for surprise attacks, without the force in the center of the village being able to defend them."

"But," the architect explodes, "without the projections I won't be able to build a hundred houses, as provided in the planning of the farm units."

"That's none of our business," say the officers.

The archaeologist, who till then had maintained a polite silence, found an opportune moment: "Gentlemen, what say we forget these three lovely hills, full of shards and holding out promise as wonderful excavation sites, and seek some other place?"

I cast an unequivocal glance at him.

"What I mean," he says apologetically, "is that if you do eventually decide on one of the hills, you should take into account that any ancient structure discovered on the hilltop would have to be fenced and preserved."

One of the settlement movement functionaries famous as a compromiser now intervenes, suggesting part jocularly part seriously :

"Gentlemen, why don't we cast lots? Here, I have a handkerchief, let's tie knots and pull."

"I also fancy hill 'B,'" says Benny. "It's finely situated in the heart of the territory, and the distances to the cultivation plots will be short. The settlers won't have any difficulty transporting their produce from the field to the sorting and consignment sheds in the village center."

I sense a sort of mute consensus, or at any rate a majority, in favor of hill 'B,' which I had liked from the very start. In an effort to "neutralize" the armed forces, I address the two officers:

"Dear friends, what's in your minds? That we should put up the settlements here in accordance with all the rules of strategy and tactics of the Israel Defense Forces? You've picked on hill 'B,' Benny and the architect have also settled for it. But what to do, there's a problem of some twenty houses deviating from your concept of a proper disposition? So they'll deviate slightly! Either you propose to the Area Command that some solution be found which is not in the books, or, failing that, in the event of trouble we vacate those houses, moving the settlers into the center of the village."

There is an indistinct growl from the Army, which I take to be a semi-mute consent.

Turning to the hydrologists and road engineers, I say: "And you, friends, will kindly draw up the necessary plans for water and access roads to hill 'B.' As to extra budgets, prepare accounts and submit them to the treasurer. A way will be found somehow to squeeze out some more money for you."

They all place their charts on the ground. Benny marks hill 'B' on his chart with a red chinagraph pencil, the others following suit, each on his own chart.

Thus is hill 'B' chosen, under favorable auspices, as a smallholders' cooperative settlement in the Lakhish Region — with the name of Nogah.

The central area of the Lakhish Region, where we wanted to place most of the immigrant *moshavim* as well as the new town, had been the scene of a decisive battle between the Israeli and Egyptian Armies during the War of Independence. Metzudat *Yoav* is located there — the former British police fortress of *Iraq Sueidan,* from which the Egyptians attacked and shelled Kibbutz Negbah, whose successful, though costly stand was no doubt the major factor in holding up the Egyptian advance northward. The fortress was later the target for repeated counter-attacks by our forces, being captured only after bitter fighting. It was in this region that the Givati Brigade fought and broke through to the beleaguered Negev settlements. Here, in the *Faluja* Pocket, the Egyptian brigade in which Abdul Nasser served was trapped and encircled.

Signs of the war were still preserved in the territory, when the planning staff's first land and soil surveyors descended upon it. The hills were full of foxholes, communication trenches and firing positions. Thousands of shell craters were still evident. Rusty barbed wire lay all over the place. We were especially concerned, however, by the countless mines and shells scattered about, particularly by the former. There was a terrifying confusion of both Egyptian and Israeli minefields.

My mind was beset by fears of disasters to our jeeps combing the territory, and to the dozens of people engaged in planning and survey, who had to walk through the fallow fields and climb every mound and hill. And what would happen once we brought in the many settlers, who hardly knew right from left?

We decided to do nothing there, not even survey work, before we had cleared the territory of mines. Easier said than done, though; no chart or information on the minefields was available from Army sources, and we had to proceed on the assumption that the entire area was mined and dangerous. Going to the top level of the Israel Defense Forces, Levi and I had to employ all our powers of persuasion before we finally secured a Sherman tank especially adapted for mine clearing. Set on the front of this tank was a "spider" — a metal cylinder with chains and cables attached; lashing the ground in front of the rolling tank, its "legs" cast up and exploded the mines uncovered by them.

The tank did the dirty work, traveling around the region for a number of weeks and blowing up many mines. At the same time, I contacted some expert sappers and got them to help out in clearing the area.

We did not make do with that. Once the sappers were through, the deep plowing of the region's soil commenced. We aimed at two targets in this plowing; first, to make sure no mines whatsoever remained in the ground; second, to furrow the virgin soil, which had lain desolate and uncultivated for centuries.

Enormous yellow tractors, "D-8s" and "D-9s," started working the length and breadth of the plains of Lakhish, dragging giant knife-plows whose blades penetrated half a yard and even deeper, leaving a trail of huge clods in their wake.

Once the deep plowing was over, we knew that this soil was ready to accept the seeds of Abraham : Jews who were also due to undergo a sort of deep plowing.

## CHAPTER XIV

## "IT'S A BOY!"

Pressure of circumstances — not calculated planning — dictated the establishment of two large *ma'abarot* in the north-eastern part of the Region ; there simply were no other absorption facilities in the Region so far. All *ma'abarot* in the country were over-crowded, and immigration continued in waves, mainly from Morocco. Working round the clock, we set up two large *ma'abarot* in the Region's northern fringes, so as not to interfere in our planning work : Haruvit, a few miles north of Kibbutz Kfar Menahem ; and Massuah, about seven miles east of its sister *ma'abara*.

Haruvit was the larger of the two, with a planned absorption capacity of about a thousand families. Rows of silvery tin huts and prefabricated communal structures very soon sprouted all over. A dirt track led to it from the end of the narrow road to Kfar Menahem.

In its early days, Haruvit suffered from the absence of health fund services. It had, admittedly, a qualified nurse, who treated hundreds of cases day and night ; but there was no clinic as yet, no doctor, no equipment, no forms or stamps. The poor nurse was never short of work, and what she could not cope with she sent to the nearest hospital, the Kaplan Hospital near Rehovoth.

At first the hospital admission clerks were forgiving in

their attitude to our immigrants, who arrived at the hospital without forms or appropriate certificates. Each time they were told: next time we will not accept you in the casualty ward if you do not bring the correct form.

In time, the clerks received instructions from higher up to be stricter with the people arriving from Haruvit and Massuah, not letting them into hospital unless they brought forms. Our volunteer instructors started driving to hospital every morning with a pickup truck full of patients, taking up the argument with the clerks. These boys were hot tempered and they almost came to blows with the admission clerks. The latter contended : "We're merely following orders. We also regret having to turn the people away, but we received strict instructions."

I immediately convened a joint conference with members of *Kupat Holim*'s management committee, the Ministry of Health's administrative staff and other bodies.

As is usually the case at such inter-office meetings, the buck is passed with everybody beating the other man's breast.

The *Kupat Holim* and the Ministry of Health people blamed each other, both sides blaming the Lakhish Region's technical department for not meeting the timetable for the establishment of a clinic. Everybody blamed the Public Works Department for not laying a road to the place, and the Ministry of Finance for not allocating funds.

I suggested adjourning the conference, and that we meet in a smaller quorum at Haruvit itself, where we would consider on the spot what should be done.

On the appointed morning, a Ministry of Health district medical officer and nurse, with their opposite numbers from *Kupat Holim*, arrived at our office in Ashkelon. The four knew each other well, after years of common work and conflict.

## "It's a Boy!"

The nurses were real old-timers — from among those, apparently, who had administered immunization injections to Harzfeld, Eshkol and Golda in the days of the Gdud Ha'avoda on the old Tiberias-Zemah road, and had pricked the posteriors of the infants Moshe Dayan at *Um Junni* and Yigal Allon at *Mes'ha*.

On that morning, the windows of heaven opened, and a miniature flood descended on the Lakhish Region. No one voicing doubts or regrets, Levi Argov and I switched on the ignitions of the two best jeep station wagons in the Region, and off we went.

Somehow we got to Kfar Menahem. From there to the *ma'abara*, though, we had to maneuver continually with all the jeep's special gears, sliding and getting stuck in the mud. We were wet through and through by the time we reached the clinic shack, where the nurse and some instructors were awaiting us.

Their worried faces told me something had happened. Addressing the doctors and matrons, the nurse said :

"One of the women was this morning taken with very difficult labor pains. She's out there writhing in pain in her tin hut, and all my efforts to ease the birth have been in vain."

The doctors and nurses set out posthaste for the hut. We stayed in the clinic shack with the instructors, who made us tea and chatted with us. The rain poured and poured nonstop. After half an hour, two of our "medical mission" returned.

They told us briefly that the pregnant woman, Sara Mizrahi, was in a very difficult state. She was losing blood fast, and if she did not receive urgent hospital care — both her life and the unborn child's were in danger. Only an obstetrician and operating theater could help in this instance.

We all rushed to the hut. We removed the woman on an

105

improvised stretcher and placed her in Levi's station wagon with the two matrons. Taking the two doctors in my station wagon, we made a dash for it.

After only a few yards, the dash became a desperate struggle with the deep mess of mud which had once been a dirt track. We advanced at a snail's pace, every now and again turning aside from the curbstones. At the exit from the camp we had to go through a deep puddle, a veritable lake, with no possibility of assessing its depth. I was in the lead and somehow succeeded in gaining solid ground on the other side ; Levi's jeep, though, got stuck in the middle. The doctors and I got down into the lake, treading in knee-high mud. The instructors and some of the inhabitants also came running. We tried to push the jeep forward, the while the woman lay inside vainly screaming to high Heaven.

The jeep settled on its belly and axles, like a stubborn mule. Removing the woman, we carried her to my vehicle, the doctor and nurse joining me.

By the grace of the Matriarch Rachel, whose own labor pains are recorded in the Bible, we reached Kfar Menahem's asphalt road after half an hour of very tough driving. Until I reached the road, I neither heard nor wished to hear anything; I concentrated on wheel and gears. Once I got onto the asphalt road I asked the matron, who was seated behind with the woman, how she fared.

"Very bad," came the answer. "If you can drive faster now, do so; we may yet save her."

The pregnant woman was half-sitting, half-reclining on the station wagon's rear seat.

She no longer screamed; her strength was ebbing; only prolonged moans and muted whimpers escaped her lips.

I opened up my engine on the narrow and winding asphalt road. When we reached the main road, toward the Masmiye junction, I switched the lamps full on, driving fast and blowing the klaxon non-stop.

We finally arrived at the hospital entrance. The woman was instantly transferred to a stretcher, disappearing with her entourage into the hospital's corridor.

I was worn out, tired and nervous. Back at Ashkelon, I showered and changed clothes.

Returning to my office, I asked Shella to phone the hospital. A few minutes later Shella came in, announcing: "Congratulations! It's a boy. Mother and child are doing well."

At the next regional conference we toasted the health of mother and son.

## CHAPTER XV

# VISIT OF THE OLD LADY

One winter day, when we were at the height of our planning work on the large *ma'abara* of Haruvit, I received an urgent telephone call from Jerusalem, from the Ministry for Foreign Affairs.

The Director of the North American Desk was on the line. "Listen," he said, "Mrs. Eleanor Roosevelt is arriving in Israel in a week's time, and we'd like you to be her host for one day in Lakhish. Show her around the project and explain it to her."

"What project?" I asked, all innocence.

"Lakhish, of course," he answered.

"Don't make me laugh, friend," said I. "We have nothing to show as yet, other than a few improvised office structures, some confused charts, and a lot of mined and thorn-covered fallow fields. Leave me alone. There's quite enough to show in Israel besides Lakhish, and there's no need to trouble Mrs. Roosevelt to come down here."

"You don't understand," said the Foreign Ministry man. "In America they've already sold her Lakhish, and got it into her head that it's the pearl of Israel's planning today. She'll be offended and hurt if we don't take her to Lakhish."

I told the fellow what I thought of him and his colleagues for selling unmanufactured goods. I added a piece of my mind about the Foreign Ministry and the Government in general, and then finally capitulated.

"I understand that during the early morning she's visiting the 'Yarkon-Negev' waterworks at Rosh Ha'ayin. I'll await her at the Masmiye junction at ten o'clock, conduct her on a tour of the Haruvit *ma'abara* and the neighborhood, and take her to lunch at Ashkelon."

"No, no! What are you talking about?" Foreign Ministry exclaimed with alarm. "You'll join General Ben-Artzi, Director of *Mekoroth*, and Mrs. Roosevelt at 7 a.m., accompany them to the Yarkon works, and then take her to Lakhish."

"Whatever for?" I asked. "Pity to waste a few working hours of mine."

"How do you mean, pity?" the fellow wondered. "After all, we can't let her travel a whole hour, from Rosh Ha'ayin to Masmiye, without company or explanations. We simply mustn't leave her alone on the way."

On a gray and rainy morning General Ben-Artzi and I presented ourselves at the entrance to the Ramat Aviv Hotel, then the pride of Tel Aviv's hotels, located on the northern bank of the Yarkon.

Eleanor Roosevelt, with her two female companions, came out at seven. We introduced ourselves and she, who had seen innumerable generals and project directors in her lifetime, looked down at us from the height of fully six feet — neither of us is particularly tall — with a good-natured smile on her face. It was a special kind of face — not good-looking, yet possessing a charm mingled with nobility, the wisdom of years and endless curiosity.

We got into a long, black car. Ben-Artzi started explaining to Mrs. Roosevelt that we were going to see a river being harnessed, so that its waters could be made to flow to the Negev.

"And what's the name of the river?" Mrs. Roosevelt asked.

"The Yarkon River," said Ben-Artzi, adding hastily : "Here — this very minute we're crossing the bridge over the river."

Hardly had he finished, when in an instant our car sprinted across the small Bailey bridge which then spanned the Yarkon, speeding on its way.

Stretching her long neck around, Eleanor Roosevelt peered through the window and asked Ben-Artzi with great wonder: "What river? What bridge?"

Arriving at the sources of the Yarkon at Rosh Ha'ayin, we stepped out of the car and entered the large tunnel which had been dug in the side of one of the hills. This was where, for reasons of security, the motors and pumps had been installed — the equipment for pumping the waters of the Yarkon into the large-diameter pipes, through which the water was to flow to the Negev.

Ear-splitting noise and a mighty tumult greeted us inside the tunnel. Welders' oxyacetylene blowpipes emitting sparks, the row from fitters' tools, motor assemblers dashing round their engines and screaming instructions and advice at each other, concrete pourers operating the mixers.

Eleanor Roosevelt looked at all this by the light of the electric bulbs and searchlights and through the rising curtain of dust. She was complimentary: "Grand," "Wonderful," "Very exciting." We then went outside to show her the actual sources of the Yarkon, the springs and the manner whereby they were caught by means of large concrete siphons. There was a matter of a few hundred yards of boggy mud separating pump tunnel and springs, and Mrs. Roosevelt hopped along in our footsteps like a long-legged stork, tramping through the sticky mud.

We finally reached the concrete siphons sunk in the ground, cylindrical in shape. Ben-Artzi requested Mrs. Roosevelt to lean over the concrete balustrade and peer at the bottom, and she did as he asked. He explained that at the bottom of each such cylinder a number of flowing springs were now harnessed, and instead of the water going to waste in the sea, it was directed to the pumps and from there directed to the Negev.

We all peered inside, and at the bottom of the cylinder, at a depth of over ten yards, we observed a tiny spring bubbling, its trickling water making a tinkling sound.

Eleanor Roosevelt looked at the small spring for a few minutes, then glanced at us as though not quite sure of our earnestness, finally asking: "Do you mean to tell me that all those giant engines I saw in the bowels of the earth, were installed to deal with that little water flowing down below here?"

Explaining that this was what we had in our tiny country, we suggested it would be futile for her to make comparisons with the Tennessee Valley Authority, the Grand Coulee Dam and her country's other giant projects.

The hour was close to nine, and the tour of the waterworks concluded. The great lady appeared tired. Parting from Ben-Artzi and his entourage, she got back into the long, black car. From there on the guest was my responsibility, and I bore in mind the Foreign Ministry's injunction not to leave her uninformed for even one hour.

Getting in beside the driver, I glanced back and observed she was tired and drowsy. I thought to myself: what harm has this Grand Old Lady done me or the Jewish people that I should now bother her with boring facts and figures; why not let her doze?

Without uttering a word, I looked to the front. Turning my head five minutes later, I noticed Mrs. Roosevelt reclining in her seat, sleeping the sleep of the just. I did not disturb her rest until we reached the entrance to the Haruvit *ma'abara*.

111

When I was told of Mrs. Roosevelt's visit, I had debated what I could show her : Charts ? Aerial photographs ? Architects' plans ? What possible interest could she find in all that ?

The only live product was the *ma'abara* with its thousand North African families, all tin-hut dwellers. I decided to bring Mrs. Roosevelt to the *ma'abara*. And in order to invest her visit with as festive a nature as possible, I requested the volunteer kindergartens and teachers to have the kids dressed in their holiday best, line them up in two rows at the entrance to the *ma'abara* and teach them to call out "Welcome!" I also requested that some of the children present the guest with a bouquet of flowers.

What did the charming kindergartners do ? They set out with the kids that morning to the nearby hills decked with wild flowers, picked handfuls of yellow chrysanthemum, mud-covered roots and all, and prepared dozens of such bouquets, the gross weight of each of which was a pound at least.

As Eleanor Roosevelt and I descended from the car, there stood the two rows of boys and girls in uniform blue, holding yellow bouquets.

The Old Lady and I started walking between the two rows, when, at the command of the head kindergartner, the children commenced throwing flowers at us. The kindergartner had apparently concluded that this was the custom in America, and wanted to provide the guest with a homey atmosphere; unfortunately, with the flowers came also the lumps of earth stuck to their roots.

Eleanor Roosevelt was stupified by the loving shower of missiles descending upon her, and her dress was spattered with mud. Blanching, I began darting about to grab the bouquets while still in flight, just so they should not hit my "Lady," all the while calling to the children and kindergartners to cease the bombardment.

## Visit of the Old Lady

All's well that ends well. The Old Lady was blessed with a healthy sense of humor. When I explained to her the secret of the reception prepared for us, she laughed heartily. She chatted with the teachers and kindergartners, hugged the tots in the kindergarten, and heard from me what the plans were for Lakhish and how these children's parents were to be transformed into good farmers. She departed in the evening, carrying with her a very large bouquet of wild yellow flowers, sans roots and earth.

## CHAPTER XVI

# PANGS OF PARENTHOOD

By the spring of 1955, the first of our settlements, Otzem, was ready for human occupation.

The painted dwelling huts were up, with an asbestos W.C. and a faucet next to each one of them. In the center of the settlement were the communal huts : cooperative store, synagogue, ritual bath, club, armory, kindergarten and village school.

Settlement on the land of the first group was set for the second half of May.

In the large and teeming Haruvit *ma'abara,* emissaries of the various movements had already started forming settlement nuclei. Two trends were active there — the *moshav* movement, which for the previous three or four years had been very active and enthusiastic in doing volunteer work among the immigrants, and the settlement movement of the *Hapoel Hamizrachi* Religious Party.

Both movements pounced on the new immigrants. The heads of the *moshav* movement, which was of a fundamentally secular nature, soon realized that they had to adapt themselves to the traditional way of life of the immigrants from Middle Eastern countries, and they sought observant men from their veteran settlers to work with the immigrants. The *Hapoel Hamizrachi* settlement movement, of course, encountered no such problems.

The two movements competed energetically. Each knew that the more persons it acquired, the better were its chances to receive squares from us in which to settle its groups.

On the face of it, they were a continuation of the great tradition of pre-State settlement, when emissaries of the settlement movements traveled to the Diaspora to influence the youth to join pioneer training nuclei.

Reality was now utterly different: Jews arrived by the thousands, with their aged, their women and their children. The immigrants made no distinction whatsoever among the finer differences between the movements, gave no thought to the ideological structure of the cooperative small-holders' settlement. They knew one thing: that they had immigrated to the land of their fathers, to "sit each under his vine and under his fig tree."

The activity among these immigrants, therefore, took on a slightly different color from the persuasion and information campaigns characteristic of the old pioneer movements. The emissaries very soon realized that they had to catch the heads of the larger and more important families, while still in the immigrants' countries of origin. Once the head of the family was on their side, all the other members of the family followed him.

The immigrants, for their part, soon grasped that they were a wanted commodity and had their price.

In order to lessen the friction and competition between the two movements, we arrived at an unwritten understanding and agreement, that the Haruvit *ma'abara* would be the *moshav* movement's domain, while the Massuah *ma'abara* be left to *Hapoel Hamizrachi*.

As the actual settlement on the land grew near, the *moshav* movement emissaries became more active among the inhabitants of Haruvit. The first nucleus which came into being there was a group of Moroccans, from the village of

Bougmez in the southern Atlas Range.

They had thick black beards and eyes like glowing embers, and were dressed in brown and white galabias. The young men were strong and healthy, all with large families; the women wore colorful dresses, as was the custom of the Jews who lived among the Berber tribes. It was as if the Jews of Bougmez had been dropped into Israel from an ancient civilization. Their forefathers had lived among the Berbers in green valleys on the southern slopes of the eternal snow-covered Atlas Mountains, in picturesque villages, each surrounded by a red-brick wall and ruled by the "Qsar," the local ruler's lofty fortress.

These villages subsisted on meager husbandry, mainly date plantations; but the Berbers were also proud warriors and excellent horsemen, ever since the days of the invasions of the Iberian Peninsula. The Jews had integrated in this wild and grand landscape for countless generations, possibly since the Second Temple era. In this society they fulfilled a well-defined function: they were the cobblers, tailors, iron- and coppersmiths, silver- and goldsmiths, and, of course, the peddlers and dealers in the colorful materials and glittering jewelry in the covered bazaars of village and townlet.

The weather in the Atlas Range tends to extremes: snow and frost in winter; blazing desert heat in summer. Weather and landscape had imprinted their mark on the Jews and their apparel. Faces burnt by wind, sun and snow, there was no city softness in them: the men were real men, the women women of the old school. The household was ruled indisputably by the male; the family was large and patriarchal; the hierarchy was clear-cut and clan discipline very strong.

The *Hakham* of the community was the final authority on matters of religion and personal status.

It was on these Jews that the emissaries of our movements now descended.

116

Bezalel Nehorai, selected by the *moshav* movement to organize the Bougmez Jews, bring them to Otzem and serve as their instructor during the first years of settlement, was himself an exception among the movement's emissaries.

To start off with, he was not a *moshav* man. He was a veteran kibbutz man, one of the founders of Ginegar in the Valley of Jezreel. In his youth, in far-off Russia, he was swept up by Zionism and *halutziut*. There he also met and married Ahavah. At Ginegar, Bezalel and Ahavah were among the mainstays of the kibbutz — Ahavah with her boundless kindness, patience and motherliness, and Bezalel, for all his bearish walk, with his agility in all handiwork in the home and in the field.

Toward the end of 1954, Bezalel, who was already in his mid-fifties, decided to volunteer with Ahavah for work in the immigrant *moshavim* on behalf of the *moshav* movement. Both were among the oldest of the volunteers, but we soon realized their merits and were happy to accept them.

What was even more to the point, in their kibbutz Bezalel and Ahavah had managed to observe religious and traditional customs. The kibbutz, admittedly, was not orthodox; but the Nehorai family observed kashruth, as well as the Sabbath and feast days. There was simplicity and naturalness in the way they did it.

Everything went fine during the first weeks of their stay among the Bougmez immigrants at Haruvit. Bezalel succeeded in establishing contact with the heads of the families, took them for a tour of Otzem, and did his best to explain to them what was going on. The *moshav* movement also placed an interpreter at his disposal, a young Moroccan from Moshav Yashresh, who mastered the Berber-Arab dialect spoken by the Bougmez people.

Some time later, Bezalel sensed that a sort of evil spirit had taken hold of his "trainees." Grumbling, murmurs, whispers; every now and again his ear caught the words "it's *taref* here," "there's pork here."

Bezalel related this to Shaul Oren, chief instructor at Haruvit. A brief tracing revealed that a number of outsiders had been "planted" at Haruvit, whose task it was to "explain" to the nuclei that they were being sent to places of "defilement and ritually unclean food."

Seeing that the "wars of the Jews" had started, I summoned the representatives of the *moshav* movement and of *Hapoel Hamizrachi,* I pointed out that soon, from May onward, we would be ready to settle a new group once a week, and earnestly requested them to stick to our gentleman's agreement and not snatch people from each other. They were up in arms, accusing each other of incitement among the immigrants, and it was with great difficulty that we finally reached some sort of *modus vivendi.*

May 22nd was declared as the day of the first settlement of the land, at Moshav Otzem.

The heads of the *moshav* movement had no thought of foregoing so great a day. The Lakhish Region was already in fashion, with headlines devoted to it in the newspapers. My colleagues and I in the Region had no great desire for mass celebrations. For us it was a beginning, and we sensed the full weight of the many gray days to follow. We also had misgivings over the behavior of the leaders of the Bougmez community, due to settle at Otzem.

One evening, shortly before settlement day, a number of *yeshiva* students arrived at the Haruvit camp. Taking up positions in front of the hut housing the volunteer instructresses and nurses, they commenced chanting "Prostitutes" and "Shameless," until driven off ignominiously by the instructors. There was no way of knowing who had a hand in this demonstration, but our concern deepened.

Bezalel had the settlers draw lots for the dwelling huts.

The huts were attached to the plots of land, in accordance with the system which had taken hold among our planners, and the village was rather dispersed. Naturally, some of the huts were closer to the village center — viz. the cooperative store, the synagogue and the other communal buildings — and others were at some distance from it. That was why we preferred to have the settlers draw lots, so as to avoid claims of discrimination. Bezalel reported that the draw had proceeded peacefully, and that every settler knew where he was going to live.

Preparations for the settlement festival were completed. A platform with loudspeakers had been erected in the village center, flagpoles had been put up, and refreshments laid on for the celebrants.

The settlers were brought to their houses the day before the festival. I was standing in the center of the *moshav,* together with a group of instructors and members of the Region's personnel, when the trucks arrived from Haruvit, led by Bezalel. The new settlers descended from the trucks, and we were supposed to direct each family to the hut which had fallen to its lot. What naïvité!

Yifrah Massoud, hefty and cunning, head of a large clan encompassing some ten families, was first off the truck. He grabbed the hut nearest the cooperative store, and started propelling other members of the clan, one after the other, into the nearby huts. Bezalel shouted at the top of his voice : "Yifrah Massoud, what are you doing, this is not your hut !"

Before we could bat an eyelid, there was the head of the Ben-Hamou clan — also some ten families — grabbing the hut nearest the synogogue, propelling his little "tribe" and invading the nearby huts.

Seeing that trouble was upon them and their huts had been snatched, the head of the Ravivo clan and his family started complaining of injury. The Cohen family exchanged blows with the Yifrah Massouds; the Assoulins started a fight with the Ben-Hamous.

Bezalel ran about like mad. He shouted, requested, begged, threatened — to no avail. They were all at each other's throats, and the entire village was in turmoil.

All of us entered into the thick of things, separated the combatants, and rearranged distribution of the village — not by lot, but according to clans. Only at the day's end, when dark descended on the raging village and the people grew tired, were we able to control the tumult and get most of the settlers into the huts. Some of the families had invaded the communal buildings, refusing to budge. Everybody was grumbling and irritable ; no one was satisfied with his lot.

That was the beginning. Returning to Ashkelon at a late hour, I immediately phoned the heads of the *moshav* movement, informing them unequivocally that the festival was postponed, and that the newspapers and radio were to be advised accordingly that very night. All we needed was a festival !

Early the following morning I hurried to Otzem. I had a vague feeling that the troubles were not over yet.

I was greeted by Bezalel, who told me that during the night a "delegation" of the settlers — about a dozen in number — had departed for an unknown destination. He feared that they had gone to demonstrate and riot in front of the public offices in Tel Aviv or Jerusalem.

"Did Yifrah Massoud set out with them?" I asked.

"No," said Bezalel, "Yifrah Massoud is organizing the local revolt. He's inciting the people not to go out to work this morning."

Meanwhile the Jewish National Fund overseer had arrived to collect the settlers for afforestation and planting work in the neighborhood. This was the only agricultural work with which we could provide them for the time being.

Yifrah Massoud came out of the doorway of one of the huts, followed by about a score of agitated people. Addressing us, Yifrah said : "The people of this *moshav* will not go out to work for the Jewish National Fund, since the wages are low and the work difficult. And anyway, this *moshav*, Otzem, will shorty be taken over by *Hapoel Hamizrachi*, and will cease being a place of unclean food and of defilement."

The J.N.F. man, who was acquainted and experienced with the likes of Yifrah, set about collecting other people from the camp's periphery, and these started climbing onto the truck. Seeing this, Yifrah grabbed hold of a spade and charged the instructor, yelling : "I'll crack your head open !"

This led to fighting, tumult and confusion. The settlers who wanted to work took fright and got off the truck.

Rushing off in my jeep to the Ashkelon police station, I returned with a sergeant and some policemen. They arrested Yifrah Massoud on a charge of attacking the instructor, and led him off to Ashkelon.

At noon we heard that the delegation from Otzem had reached the institutions in Tel Aviv, and was striking outside *Hapoel Hamizrachi* headquarters in protest against "religious coercion," "unclean food" and "pork" at Otzem.

Photographers and newsmen went to town with the story, and write-ups with pictures of our Moroccans — galabias, complaints and all — appeared in all newspapers.

Of this much I was sure : if the matter were not decided immediately, there would be a chain reaction and general deterioration would result.

Proceeding to Jerusalem, I put the matter to Eshkol. He immediately perceived the gravity of the situation, and summoned all the top politicians of *Hapoel Hamizrachi* and the *moshav* movement. It was a tough and bitter session. I used very strong language in describing the affair. The *Hapoel*

*Hamizrachi* people were on the defensive and accused me and my colleagues of the "unwarranted arrest of their comrade, Yifrah Massoud." I retorted that when threats were uttered of "cracking open the heads" of our instructors, there was only one "comrade" as far as I was concerned, and that was the Israel Police.

I further told them that if Yifrah Massoud and his dozen friends were "comrades" of theirs, they were welcome to remove them from Otzem, and from Lakhish in general.

Since we are all "robbers," and in view of the fact that till then I had maintained excellent relations with the *Hapoel Hamizrachi* people—the majority of them being hard-working men — by the end of the meeting we reached a compromise. All were forgiven.

But there was no forgiving Yifrah Massoud, not until he had left Otzem of his own free will, settling with some half dozen other families in the first *Hapoel Hamizrachi moshav* in the Lakhish Region, Moshav Eitan. There Yifrah Massoud turned into a peaceful and industrious settler, and from time to time I would drop in on him for a drink and to reminisce about the fiery days of long ago.

CHAPTER XVII

# RURAL JEWS AND URBAN JEWS

The Atlas Jews who populated Otzem, the first of our settlements, were, on the face of it, rural Jews; we even thought they were "farmers," and believed that their striking roots in the soil would not involve particular pangs of absorption. We soon discovered how wrong we were. Not only is there no similiarity between modern farming and that practiced among the Berber tribes; what was more to the point was that they had never been farmers — they had been craftsmen, artisans, petty merchants and so forth. Indeed, when a while later an emissary friend of mine visited the townlets and villages of the Southern Atlas, and, going through the bazaars, sought among the artisans' stands for shoes peculiar to the Atlas, he more than once was told : "Ah, sir, there are no good cobblers now in our town. There used to be excellent cobblers here, Jews, you know; but these departed from here for the land of their fathers, and now there are none left to make fine sandals. Good galabias are also difficult to get nowadays, for the Jewish tailors, too, have departed from here and wandered eastward, to Israel..."

In this, at least, the former Atlas Moroccans did not disappoint us : even if they were strangers to modern farming — by virtue of the heritage they brought with them from Morocco, they were not strangers to handicrafts or to a rural form of life.

123

When it came the turn of Moshav Shahar to be settled, we requested the Absorption Department personnel who accompanied the immigrants on the sea voyage from Marseilles to Haifa, to prepare for us a further nucleus of Atlas Jews for settlement at Shahar. "It will be all right," they said; having so much on their minds, however, they did not bother too much in the selection of people. Coming to Shahar the day after the arrival direct from the ship of the first group of settlers, I noticed the fallen countenances of our instructors.

"What's up?" I asked.

"Come and see for yourself," they answered, and took me along to meet the new settlers.

At the first hut, we met the Dahan family. The head of the household sat in the entrance to the hut, utterly dumbfounded at what was going on around him, while the woman of the house — good-looking and heavyset — was busy with the laundry in the yard. The children, five or six in number, sat in the bare yard, strangers to their surroundings.

"Where are you from, Mr. Dahan?"

"From Tangier, sir."

"And what did you do in Tangier?"

"I had a café there," was the reply. "Is there no *ville* (town) in the neighborhood, that you brought me to a *village*? I'm no farmer, nor the son of one. I want to open a café in one of the towns."

My answer, for whatever it was worth, of course, could not satisfy him. We moved to the second hut, where we found the Levi family.

"Where are you from, Mr. Levi?"

"From Tangier."

"What did you do in Tangier?"

"I had a café there."

The third hut was occupied by the Cohen family, also from Tangier, where Mr. Cohen had been the owner of a cabaret.

The Ben-Haroush family, in the fourth hut, had had a café in Tangier.

Mr. Ben-Simhon of Tangier had owned a wine and spirits store, and had been a ship's chandler in that port town. Mr. Yifrah had been the owner of a café.

That was a cross-section of the settlers of Shahar. Before long, the former café and cabaret owners from Tangier started laying pipes, helping the builders, working in afforestation, thereafter also cultivating their own vegetable patches.

It cannot be claimed to have been an easy process, or one that all of them could easily tolerate.

Weeks and months went by. The Lakhish villages were gradually filled. Then came the turn of Kiryat-Gat, the Region's capital.

The first fifty houses, two-family units, stood in two rows, ready for occupation.

A short while before, we advised the Absorption Department personnel that this time we needed town folk. It was important that the early settlers be urban dwellers possibly with experience in modern industry; there was room also for a café proprietor, as well as for some grocers. The main thing was that they should be town folk, and preferably from a large town — Casablanca, for instance.

On a certain evening, at twilight, we were standing at the approaches to Kiryat-Gat, waiting for the first group of settlers due to arrive shortly by truck from Haifa port. There were instructors and instructresses, absorption personnel, some of the Region's management staff, as well as women volunteers who would provide cookies and cold drinks by way of refreshment for the weary travelers.

All of a sudden, the first truck arrived in a cloud of dust on the dirt track leading to the "town." With a grinding of brakes, it stopped beside us.

"Is this Kiryat-Gat?" the driver asked.

"Yes, this is Kiryat-Gat," we replied.

"Right," said the driver, a hefty youngster. Jumping from his cabin, he walked to the rear of the truck, undid the tarpaulin cover, and, drawing out a single step, called out:

"Gentlemen, we've arrived. Down you come."

And there they were, descending: heavy-bearded, in brown galabias, patriarchal types right out of the Bible. The Patriarch Abraham, with his flowing white beard; the Patriarch Isaac, tall and wide-boned, gray beard sprinkled with silver; the Patriarch Jacob, black-bearded, and eyes blazing.

I looked at them aghast.

"Where do you hail from ?" I asked.

"From the village of Ourzazate (near Marrakech)," they replied. "We have come to cultivate the land of our fathers, to plow and sow and extract bread from the soil."

Such were the first urban inhabitants of Kiryat-Gat. It was not long before these folk went to work — albeit not enthusiastically — in the young industries we started establishing in the new town.

## CHAPTER XVIII

# THE GREAT ADVENTURE

A phone call from the Ministry for Foreign Affairs: "Again we have an important guest, and this time you'll surely be pleased to receive her. Dr. Margaret Mead, the world-famous anthropologist and sociologist, is arriving in Israel."

I acceded at once, and with delight. As a student of sociology at the Hebrew University of Jerusalem, I had been a faithful follower of her books and research works, which are required reading for every student of social sciences. I loved her lucid style, devoid of the ornate ultra-scientific jargon characteristic of so many sociological studies, the sole purpose of which, apparently, is to confuse the reader and student. Margaret Mead always wrote pertinently, in an appealing and fascinating style.

Now I would be able to see her face-to-face, without having to stand on superfluous ceremony. I knew that Margaret Mead was used to and experienced in difficult expeditions, along barely negotiable tracks. I had no hesitation whatever about driving her in my jeep station wagon from her Jerusalem hotel. We were allocated three days of tours.

On the first day, I set out with her along dirt tracks to the vicinity of ancient Tel-Lakhish. Standing beside the tall *tel*, the spirit of archaeology descended upon me, and I explained to her that this *tel* had both ancient and recent layers.

The ancient layers dated to the Israelite conquest and earlier, and were at least 4,000 years old. The recent layers, at the top of the mound, dated to the Byzantine era and were about 1,500 years old.

"Most interesting," said Margaret Mead politely, "and what's that group of white houses on the far horizon?"

"That's Kibbutz Beit Govrin, a relatively veteran settlement, which occupied the area before we dreamt of setting up the Region."

"And what's the age of this veteran kibbutz?" Dr. Mead asked.

"Six years," I replied.

"Fine," said Dr. Mead.

Proceeding westward, we passed by Tel-Gat. I explained while driving: "This is a *tel* on which were discovered remains of a burial ground dating to the early days of the Arab conquest, in the 8th century A.D. The archaeologists, however, are attempting to plumb the depths of the *tel*, in the hope of finding Philistine layers."

We reached the new immigrant *moshavim* in the center of the Region, near Metzudat Yoav.

"In front of you," I told Dr. Mead, "you have three new settlements. One is inhabited by Kurdish Jews, the second by Moroccan Jews, and the third by Romanian Jews. They were established only a few weeks ago." Pointing in the direction of Kibbutz Negbah, I said: "And that's a veteran kibbutz established about ten years ago."

"Mr. Eliav," Dr. Mead interrupted me with the hint of a smile in her eyes and on her lips, "perhaps you'll make up your mind once and for all what's considered 'new' here and what 'old.' Does 'new' signify a thousand years of Tel-Gat or a month of an immigrant settlement; and 'old' — is that supposed to mean ten years of a kibbutz or 5,000 years of Tel-Lakhish?"

# The Great Adventure

I said: "It's indeed difficult to measure our history and geography with a regular slide rule. It was not for nothing that Herzl, the founder of Zionism, named his book *Altneuland*."

We entered the new immigrant villages. Dr. Mead's scientific, human and intellectual curiosity knew no bounds. She talked with the immigrants, their wives, and their children. She questioned and examined instructors, instructresses, planners, architects and engineers. She visited huts, the cooperative store, armory, synagogue, and ritual bath. Nothing escaped her probing eyes, and her questions were all pertinent. The experience of dozens of years of research among tribes, communities and races, in different corners of the earth, was evident in her enquiries.

After three days of intensive touring, I returned Dr. Mead to Jerusalem. On the way, I questioned her: "Dr. Mead, please let me have your general impressions of the project you've seen, and rest assured that my colleagues and I are open to criticism. I must tell you that we ourselves are groping in the dark, trying to learn from mistakes of the past and to apply new ways and methods."

Dr. Mead looked at me with probing eyes: "Do you really want to hear my opinion?"

"Of course," I replied.

"Well, I think you're proceeding in this matter in a bad, wrong and disorganized fashion."

"So," I said, "perhaps you'll explain what you mean."

"Certainly, I'll tell you how I would have set about it. Here you are, planning the establishment of a few dozen new villages, a number of rural centers and a new town, and you propose to settle myriads of Jews in them. These Jews come from the Atlas Mountains, the towns of Czechoslovakia, the mountains of Kurdistan, from the Yemen, India, Romania and so many other lands. I've observed

129

that not only don't these people speak a common language; socially and culturally they could come from totally different planets. Their sole common denominator, and I stress sole, is that they're Jews. Furthermore, and this factor is possibly no less important, these people are not farmers at all; only individuals among them, from Kurdistan and Morocco, have engaged in primitive farming, and it would be better for them had they not known the art when they came here. And you're thinking of transforming them, in a matter of a few years, into modern farmers within a developed and complex cooperative context, which in itself is a novel experiment in our world."

"Quite correct," I said.

"Well, Mr. Eliav," Dr. Mead said, "I'd have gone about it in the following way: first, I'd have appealed on behalf of your Government, to the appropriate U.N. bodies and request them to investigate all aspects of the subject."

"The U.N. bodies?" I wondered.

"Yes, the U.N. bodies. Once you'd appealed to them in writing, they'd answer you a few months later that they were acceding to your request, and would be sending a commission for an on-the-spot preliminary study of this weighty subject. Such a commission would be composed of representatives of the Food and Agriculture Organization, the International Labor Organization, the World Health Organization, and similar bodies. The distinguished commission would request a reasonable period of time for investigating the complex subject — say, three years. At the end of three years of ramified research, the commission would request a year's extension for writing its report. At the end of the extra year, you'd receive a report — a thick volume containing hundreds of pages. At the end of the book, under 'Conclusions and Recommendations,' only one line would appear: 'It cannot be done.' "

I had gradually caught on to the fact that Dr. Mead was pulling my leg, and now, at the end of her speech, I noticed the mischievous glint in her wise eyes.

"And so, Mr. Eliav," Margaret Mead concluded, "you went your own way. You didn't call on the U.N. and its bodies, nor did you wait for the advice of sociologists and anthropologists such as myself. And a good thing, too. This is a great human adventure, and may God bless you."

That was not the end of Margaret Mead's ties with the Lakhish Region.

Some time later Margaret Mead again appeared in Lakhish, this time with her 17-year-old daughter, a lovely, smiling, girl, long-legged and flaxen-haired.

"My daughter wants to see your border settlements," the mother sighed, "and, of course, the wonderful *Nahal* soldiers."

We went out to the Lakhish outposts, Nehosha and Amatziyah.

At the end of the tour, at Amatziyah, I was witness to a fairly stormy argument between mother and daughter. I refrained from interfering, of course.

Approaching me then with her daughter, the mother said : "Isn't it true that it's very dangerous here, and that there are exchanges of fire here practically every night with infiltrators and terrorists coming from across the border ?"

Not being quite sure of the position I was supposed to take in this argument, nor fully grasping the portent of these questions in the daughter's presence, I murmured something which sounded affirmative.

"You see," said the mother to the daughter, "you can't stay here. It's absurd."

"But mother," said the girl, "there are also girl soldiers here. Why shouldn't I stay ?"

"Isn't it true that she may not stay here?" Dr. Mead appealed to me, and I discerned a note of entreaty in her voice. The anthropologist had given way to the mother.

"True," I said.

The daughter started getting annoyed and displaying signs of obduracy.

"I'm staying here, and that's that!" she told her mother. "You didn't ask anyone's permission when you went to live among the head-hunters of Borneo, and I won't ask anyone for permission now. I like the place. The boys are cute, and I'm staying."

Having no option, the mother consented to her daughter's staying the night. I placed her in the capable charge of the company commander.

"Treat her gently," I told him.

"Rely on us," he replied, winking.

Margaret Mead left Israel a day or two later. The daughter, however, stayed at Amatziyah not one night, nor two, but many weeks. I should not be surprised if, one day, I come across a book entitled *Pangs of Adolescence in the Lakhish Region*.

## CHAPTER XIX

# DEATH OF BEN-AMI

When the volunteer instructors from the veteran *mosha-vim* started arriving, I was pleased to meet among them also some of my comrades from British Army days, who had thereafter followed a path similar to mine in the *Haganah*, illegal immigration and, of course, in the War of Independence, and the Israel Defense Forces. Men my age, in their mid-thirties, once again they abandoned plow and home, leaving the family homestead only recently established after years of wandering and fighting; again they responded to the call, setting out this time to help the immigrant settlements. They went sans rhetoric, parades or farewell parties.

These men and women became the solid mainstays of the immigrant *moshavim*, being assisted by younger volunteers in their twenties. They became familiar faces in my little home in Ashkelon's Rehov Havradim — not merely by virtue of our mutual past, but also because I had only recently been a volunteer instructor at Moshav Nevatim.

Among the first to arrive in Lakhish was Ben-Ami Malkiman of Kfar Yehezkel. No speaker or leader, he yet stood out immediately by his steadfast personality. I first came across him in Italy, in 'forty-six. He had served in the Jewish transport units of the Royal Army Service Corps, and after the war had stayed on to work with Yehuda Arazi in the immigration organization in Italy. Ben-Ami was already then known as a driver par excellence and first-rate organizer of convoys, transport and sailing camps. He could do any job in our branch in Italy.

I remembered him as a young man of above average height, strong and handsome, with a kind and open countenance. His outstanding trait was his inner quiet, a quiet bordering on taciturnity. He spoke little and did more than anyone else. He had no liking for idle talk, gossip or profanity, in the manner of old soldiers, nor did he tend to join us in drinking and living it up in Milan or Naples following successful operations. It was soon evident that he was forming an attachment for one of the refugee girls working with us at the Magenta Camp.

Magenta was a small village some nineteen miles from Milan. The central supply base for our operations in Italy was set up at Magenta. There, on an isolated farm, we stored transport and fuel, navigating instruments, and at times, also arms acquired by us. It was there that the *Pal-Yam* people were accommodated. To Magena was brought also a select group from among the Jewish refugees, to assist in our work. There were a number of girls among these refugees, and with one of them — Zippora, who looked for all the world like a beautiful Slav girl — Ben-Ami fell in love. We jested and made fun of the love birds; there was also a shade of envy of these two, who built themselves a family nest within the urgent and tense work surrounding them. Back in Palestine, Ben-Ami brought Zippora to his village, Kfar Yehezkel.

He joined the Army at the outbreak of the War of Independence, and upon his demobilization set about building farm and home. For the next few years I did not see much of Zippora and Ben-Ami : I had heard that three children were born to them, and that Ben-Ami had built up a splendid farm in his village. And here he was now in Lakhish, among the volunteers. Our friendship was immediately re-established; once more I found in Ben-Ami a good partner, quiet, serious, second to none in devotion to the job.

Ben-Ami excelled in his work, and was appointed in charge of instructors in a number of villages in Lakhish and the Negev. A pickup truck was placed at his disposal, and he would travel from one village to the next, checking on the teams, seeing to their needs, and maintaining contact between them and the *moshav* movement and Settlement Department. Wherever he turned up, his constant smile, calm manner and loyal assistance produced a good atmosphere.

Toward evening one day, Ben-Ami set out from Moshav Shahar in Lakhish in the direction of the new immigrant *moshavim* near Sa'ad, on the approaches to the Western Negev.

That evening, they waited for Ben-Ami in the *moshavim* of the Western Negev. When he did not turn up, the instructors phoned the next day to ask why Ben-Ami had not arrived, setting in notion the machinery of investigation and enquiry. After fruitless phone calls to other *moshavim*, to Kfar Yehezkel and Tel Aviv—fear and worry gradually started gnawing at us. What had happened to Ben-Ami — our methodical and reliable Ben-Ami? Hour followed hour, and after half a day we notified the Police and the Army, who immediately commenced a search which spread across Lakhish and the Western Negev. We joined the searchers. The anxiety and fear became a nightmare. Forty-eight hours had elapsed since Ben-Ami's disappearance, when the Negev Police were notified that an abandoned pickup truck had been standing for the past two days at the side of the dirt track near Kibbutz Sa'ad. When the policemen arrived on the scene, they found Ben-Ami's body leaning over the steering wheel, a bullet in the back of his neck.

It was at first presumed that he had been shot by *fedayeen*. But the bullet had been fired from too short a range, inches in fact, so that it became clear the murder had been committed by a person seated in the truck.

A hitchhiker? It was our custom not to refuse to take hitchhikers on our trips. There were many dirt tracks and few roads in the vicinity, and even on the roads, traffic was slight during the afternoon and evening. And if we, the Region's personnel, would not take settler-hitchhikers — who would?

Police investigators swooped down on the settlements, locating Ben-Ami's last hitchhiker. Upon investigation, the fellow admitted that he had shot and killed Ben-Ami with his rifle. The man, a young immigrant, was no different from the many immigrants Ben-Ami was always wont to transport in his pickup truck from one settlement to the next. It later transpired that this hitchhiker had run amok, and in the frenzy of his attack he fired at Ben-Ami's head, killing him on the spot.

We laid Ben-Ami to rest in his village. The bier was followed by all members of the *moshav* — the founding generation, bent by years and toil, Ben-Ami's generation, and his children's young friends. We were at a loss for words to console Zippora and the boys; we sat there mute and silent.

The next day we returned to our work.

During the following days I sensed tension among the personnel, our drivers in particular. They stopped taking hitchhikers from among the settlers. I heard, and later saw, how our cars speeded through villages and along dirt tracks, leaving behind settlers waiting by the side of the road for a lift. Summoning the Region's personnel, instructors and drivers, I told them that such behavior could not go on. We were starting to treat the settlers in the same manner Jews were treated in the Diaspora. We could not visit the guilt of the one on the many.

"This is not going to be our way of life," I said, "and we shan't continue like this. Ben-Ami himself certainly would not have acted this way."

The next day, hitchhikers were again being given lifts on the roads of Lakhish.

## CHAPTER XX

## "WILD WEST"

Even before the Lakhish Region was born, there was a mood of unrest in the country, a pioneering and volunteering spirit, a young generation's yearning for "something new and unconventional." This is apparently a natural periodic manifestation, whereby the revolt against conventions becomes a matter of fashion. Ben-Gurion's departure for Sde Boker, for all that he was "the Old Man," gave added impetus to these feelings; "Here's 'the Old Man' gone to Sde Boker, and where are we?" Thus the rise in the *moshav* and kibbutz members' trend of volunteering to help the immigrant settlements, as well as the reason for attempts at unconventional forms of settlement by Israeli youth.

When we commenced working in the Region, we found a legacy there of a few such nuclei, revolving in an orbit around Ben-Gurion and a number of his intimates. These latter came to me, saying: "These youth, who don't belong to any defined settlement orbit, have been on our hands for the past few months. How about taking them under your wing?"

These groups caused us no little trouble and disappointment. There were boys and girls of all possible types and classes, who were simply tired of the boring provincial life in the veteran Sharon Plain villages from which they came.

They had not found their way to the youth movements and organized frameworks which led to kibbutz or *moshav*. Now they wanted "something to do."

There were among them eccentrics from the cities, seeking a vent for all sorts of desires. It was difficult to estimate their number and the degree of their earnestness.

We called a conference of all these groups at Ashkelon's community hall. We invited Ben-Gurion and Eshkol to the conference; Moshe Dayan, who was then Chief-of-Staff, also came. There were orations and addresses by the leaders on settlement and pioneering. Standing in front of a huge map of the Region on the stage, I told them of our plans for this region. After the conference, we held many talks with representatives of miscellaneous and varied nuclei.

On the whole, the talks would go somewhat as follows:

"How many are you?"

"About thirty."

"How many boys and how many girls?"

"Twenty boys and ten girls."

"What part of the country are you from?"

"Kfar Saba and the vicinity."

"What form of settlement were you thinking of setting up in the Region?"

To which they would respond with a question: "What forms of settlement are you setting up in the Region?"

We replied: "We're setting up immigrant *moshavim*, Israeli youth *moshavim*, collective *moshavim*, as well as kibbutzim, of course."

"Is that all?" was the reaction. "After all, these are quite conventional forms, and we thought you were doing something really new here."

"How do you mean, really new?"

"We were thinking of something urban."

"Urban?" said we. "Very well, we're also planning to

build a new town in the center of the Region; come, you
can be among the founders."

"What sort of town will it be?" they asked.

"What sort of town? A town like all towns. A small
town to start off with, eventually growing into a big town.
It depends very much, of course, on its initial settlers."

"But we don't want just any town," they told us. "We
want a town with rural characteristics, or rather, a village
of an urban nature; something agricultural which will have
industries, or something industrial with agriculture thrown
in. We're also searching for something collective, but not
of the kibbutz type; something having the cooperativeness
of the *moshav*, but neither *moshav* nor collective *moshav*;
something where we can both live as Israeli youth and help
immigrants; a framework in which we can be on our own,
and yet not on our own..."

And so on, and so forth.

We proposed to all these groups that they gather together
in an agricultural training camp we would establish for them
within the Region. We built a camp of a few dozen huts at
the Ibbim Ranch in the south-western part of the Region,
and invited them to come and work there. Of the hundreds
of members of the nuclei, a few dozen boys and girls came
to Ibbim, where they worked under conditions similar to
those of the *Nahal*.

There was a lot of traffic and no little instability at Ibbim;
finally, a group of about two dozen crystallized, deciding on
a *moshav shitufi* as its aim.

We got them attached to this special *moshav* movement,
and in due course they established the *moshav shitufi* of Nir-
Hen in the heart of the Region. They also went through
difficult times, and only a handful remained at Nir-Hen, which
in the course of years became an ordinary *moshav* which
absorbed ordinary people.

140

Some time later we encountered private initiative of a totally different kind.

One day Levi Argov brought Mukka in to see me. Then aged thirty, Mukka was tall and broad-shouldered, with coal-black hair and eyes like embers; hawk nose above a magnificent Terry Thomas mustache, bronzed face and muscular body. Mukka was native-born, and raised in one of the settlements in the South. His father was from Georgia in the Caucasus, a member of that tough and wonderful Jewish tribe, whose origins I got to know a few years later when I served at the Israel Embassy in Moscow.

All I knew about the Caucasian Jews at the time was that they had provided some of the most picturesque members of the *Hashomer* organization — with their Circassian costumes, large cartridge belts and Astrakhan caps, riding their thoroughbred mares.

Mukka, second-generation Israeli of this stock, had come to Lakhish as a settlement organizer.

And what was Mukka's idea? Not for him or his likes settling in the cream of the Region, in the plain, and raising cotton and vegetables. Only the hilly terrain fronting the Hebron border would suit him. There, in the Region's "Texas," he and his friends would establish a settlement based entirely on cattle raising; a settlement of horse-riding cowboys, who would graze their herds in the vast pasturelands, and incidentally having the time of their lives.

There was a special charm about Mukka. I requested Levi to study him closely, and to meet with his friends.

After meeting with a group of about twenty boys and a number of girls, Levi reported that although there were some peculiar aspects to them, they had certain advantages over the previous nuclei. In the first place, most of them were farmers and of farming stock; secondly, they wished to settle in a location with security hazards.

I met once or twice more with Mukka and his group, and I must confess, I was infected by their romantic spirit. I wanted to help them realize their ambition.

Benny Kaplan remained skeptical about them. He warned me repeatedly that we would be backing a dubious venture by placing a land "square" at their disposal. He was in the minority, however, and had no option but to locate a site for them.

In fine hilly terrain in the north-eastern part of the Region, north of Kibbutz Gal-On, the site was located for the new settlement of Beit-Nir.

We built Mukka and his people about ten dwelling huts, a dining room and some shacks; office, armory and the like. We also set the date for celebrating the "day of settlement of the land." That day we had news of terrorist and *fedayeen* attacks in the south, and the journey in three jeeps from Ashkelon to Beit-Nir was very strained, our weapons on the ready.

Captain "Zvingi," who was in charge of area defense for our settlements, escorted us in a patrol jeep. Zvingi, a member of Revadim in the Etzion Bloc and a veteran fighter, had been among the founders of the new Revadim in the South after the War of Independence. The four settlements of the Etzion Bloc in the Hebron hill region had fallen to the Jordan Legion on the eve of the Declaration of Independence, on May 14, 1948, after withstanding a long siege and numerous attacks. Excellent relations obtained between us and Zvingi, a born soldier and fighter. He was tall, muscular, red-haired (thus his nickname, which was a combination of "Zvi" and "ginger"), with a huge mustache — red, of course — adorning his thin, sharp face.

Drawing near to Beit-Nir, we heard volleys of shots ahead of us. It was twilight, and we found it difficult to imagine that the *fedayeen* would attack a settled area in daylight. All

the same, Zvingi stepped on the gas and his men held their
weapons at the ready. With us speeding behind, we all
rushed into the settlement.

Mukka and his comrades gave us a warm welcome. Asked
what the shooting was, they replied they had fired in the air
as part of the celebration. Zvingi went red with anger : they
had better not start their relations with the Army in this
fashion.

We visited the dwelling huts. The boys had transformed
the place into a setting taken from the Wild West. On the
dining room they had hung a large "saloon" sign ; one of
the huts had "Bar" on it, another said "Sheriff"; the girls' hut
had "Can-Can" on it, and so on.

About thirty fellows had collected around Mukka, most
of them villagers on the lookout for a bit of "wild life" and
adventure. Some were honestly keen on setting up farms,
others were plain beatniks in search of "living space" away
from any framework or law.

The girls were also different. Some were flappers from
Tel Aviv, who had heard tell of hot-blooded cowboys; others
were possessed of a pioneering spirit and nostalgia for the
"good old days"; yet others were plain eccentrics. Even Mukka was unable to tell who lived with whom.

A few months passed, and it became clear that only a handful of fellows around Mukka were really working and carrying the burden. Around them swarmed all kinds of strange
types, who came, ate, slept and went away without putting
in even one day's work. There were fights and friction between Mukka and his assistants, and the loafers and mischief-
makers.

We tried sending them instructors, *moshav* and kibbutz
people, but they were rejected. We tried to get this band to
join one of the settlement movements, the Farmers' Associa-
tion preferably, but that attempt also failed. When, after

a while, heavy financial deficits started accumulating, we called in Mukka and advised him to pack up and leave. He took our advice, and within a short space of time the settlement was emptied of the strange band.

The place did not stay empty long. A nucleus of the *Hashomer Hatzair* youth movement, fine youngsters full of energy, established a kibbutz there. And this kibbutz became one of the most successful settlements in the Region.

CHAPTER XXI

# THE GUBERS

Shortly after we had settled down in Ashkelon, Shella informed me that Rivka and Mordechai Guber wished to talk to me. I had a superficial acquaintance with the Gubers, and I guessed what had brought them to see me.

The Gubers were a unique couple, a pioneer family from Russia, straight out of some Zionist legend. In the 'twenties they had been among the founders of Kfar Bilu near Rehovoth. They were still possessed of that universal spirit of frontier pioneers and sought a new beginning elsewhere. That was how they came to join the founders of Kfar Warburg, where they cultivated their garden and raised their two gifted sons — Ephraim and Zvi — and their daughter Haya'le. There, too, however, they would not rest on their laurels. During World War II the eldest son joined the British Army; Rivka, then about forty, also volunteered, serving in the A.T.S. At end of war, mother and son returned to the farm. Then came the War of Independence. One after the other, both sons fell in battle with the invading Egyptian forces.

The period of mourning and bereavement over, Rivka submerged herself in collecting and publishing the writings of her talented sons. Ben-Gurion took her to his heart, seeing in her a paragon of the Mother of Israel. Mordechai devoted himself to immigrant absorption, finding his sphere of activity in the huge immigrant *ma'abara* at Kastina, which sprawled in the vicinity of Kfar Warburg.

It was one of the more difficult *ma'abarot*. The immigrants, mostly Kurds and Iraqis, were unhappy, stubborn and bitter as gall. Lean and short of stature, seemingly fragile, Mordechai took charge of the staff of the *ma'abara*, which was on the verge of an explosion, and where strikes, outbreaks, riots and complaints were a daily occurrence.

Mordechai's external appearance was misleading: at first glance one might think of him as a typical old-time teacher of Hebrew literature — sparse hair, high forehead, deep blue eyes, sharp nose; the only thing missing to complete the picture was a pince-nez. A second and third glance, though, revealed a different person. There was a deep grayish tan about his small features, a tan deriving from years of toil in sun and wind. The sun and wind had moulded and carved Mordechai's face, and the wrinkles told the tale of those years like the rings of a tree trunk.

And then one would notice Mordechai's hands. These hands, whose owner had forced them to handle hoe and spade, day in, day out, year in, year out, had grown apart from the body as though in protest. They were the hands of a six-foot-tall farmer, not the hands of a short-statured person like Mordechai.

These hands must have often deterred those who forced their way into his room at the Kastina *ma'abara*, threatening violence. The violent types at the *ma'abara* presumably sensed that this no-longer young and little person could suddenly throw them out of the room.

A fourth and fifth glance, though, showed that the first glance was not entirely misleading. There was, in fact, the teacher in Mordechai Guber. In his youth he had been a teacher and bookworm, and such he had remained.

It was not easy to understand Mordechai's character. Rivka's character, on the other hand, was easy to analyze. She was like an open book, and from the outset would tell one about herself, her origin, her doings, desires and ambitions.

Short and round-faced, she was tubby and ball-like. And in this ball were extraordinary wells of energy, which burst forth not only in her speech but also in her manner and her deeds; in fact, in anything to which she set her hands.

To me she seemed like a wonderful combination of the Yiddishe Mamma and the revolutionary "narodnik." Contributing to the Russian image were her long silver-black hair done in the "babushka" style, the heavy Russian accent of her Hebrew, as well as her wide knowledge of Russian literature, and her propensity to quote from it.

Rivka was the spokesman, and she explained what brought them to me.

"In fact," said Rivka, "we've already made up our minds to sell our property and house at Kfar Warburg, and move to Lakhish to work with the immigrants. We've come to you now, to find out what we're going to do here and where we're going to live."

I did not attempt to dissuade them. I made no effort to tell them that conditions would be very difficult, and that they were no longer young. If these two had set their minds on something, nobody could make them budge.

We assigned Rivka and Mordechai as instructors in Moshav Nogah. Established after Otzem and Shahar, Moshav Nogah at the time comprised a few dozen wooden huts dispersed over two low and gentle hills. The one-room shacks, which had found their way to us after seeing service in the first *ma'abarot* near Tel Aviv, were painted in vivid colors — green, red, blue. It must be admitted that by this simple stratagem we succeeded not only in masking their ugliness, but also in giving the entire village a colorful and gay appearance — with a minimal outlay.

To Moshav Nogah we brought a nucleus of Jews from Iranian Kurdistan, who had been residing for about a year in the blazing tin huts of the Tiberias *ma'abara*.

At the height of the building activity, we brought along Yehuda Sharifi and some clan leaders from the Tiberias *ma'abara*, to see the place being offered them. We went overboard in painting an imaginative picture of the future, the houses, the cultivated fields, the plantations.

Yehuda Sharifi, broad-shouldered and of average height, had been a tailor in one of the little townlets in the mountains of Kurdistan. He and his comrades admittedly knew nothing about modern farming, but they were used to a rural landscape. They looked about them at the bare hills, and agreed immediately to come and settle at Nogah.

Since they were few in number, in our anthropological innocence we matched them with a further nucleus — of Jews from Iraqi Kurdistan. From the very first day, this forced integration led to friction and quarrels between the two communities, which differed one from the other in both language and customs. It took years for the differences to blur somewhat.

To this village we assigned Rivka and Mordechai Guber. We believed it would not be difficult for Mordechai, who had already acquired experience among Kurdish Jews at the Kastina *ma'abara*, to find his way among the inhabitants of Nogah. A weightier consideration, however, involved the plan to put up, linked to Nogah, the structures of the Region's first rural center — Nehorah. To our mind, Mordechai was a candidate for coordinating the regional council, eventually to be set up at Nehorah.

The Gubers took possession in the spring of 1955, together with the Kurdish immigrants from the Tiberias *ma'abara*. A greenish one-room hut, neither larger nor smaller than those of the other settlers, was placed at their disposal.

## The Gubers

Within days the Gubers' hut was transformed, as by a magic wand, into a clean and warm home. White curtains on windows, and a freshly ironed cloth on the round table; bookladen shelves, and the sons' portraits on the masonite wall.

The hut was full of people from the outset, day and night. Anybody passing through Nogah or working there found it an obligation to call on the Gubers. A cold and refreshing drink awaited all comers, as well as a taste of Rivka's cookies, ever-present in the tiny pantry.

Mordechai ran local affairs, and Rivka became the schoolteacher. The school at Nogah was also accommodated in a little hut, with second-hand furnishings and improvised materials. Rivka taught Nogah's first children in this hut. There was, of course, no possibility of having separate classes. Instruction was basic and improvised but it was no mean achievement to have a school at Nogah from the outset. And Rivka had a craze : flowers. She soon succeeded in talking our water engineer into having some piping installed, and Mordechai was persuaded to allot one of the large plows to prepare a plot of land for her next to the school.

She secured seeds and saplings, and within a short while the children were hoeing and weeding among the school's flowerbeds. It was the first strip of green on the soil of Lakhish, a sort of harbinger.

No few disappointments attended the Gubers as well. They found it difficult to communicate with the depressed, heavy-footed and extremely stubborn Kurdish Jews. Language barriers also intruded.

There were tough conflicts at Nogah : "Iraqis" against "Iranians," clan against clan, and all together against the instructors. Many factors were involved : the difficult and unusual work — construction, laying water pipes, road-laying, wood planting; the tension of night watch; the occasional delay in wages; and above all — the settlers' lack of understanding and trust of the attempt at organizing their lives and future for them.

There were strikes at Nogah, as well as demonstrations and some people left. With all of this the Gubers managed despite the difficulty of such a struggle, which is, at times, almost beyond human endurance.

Meanwhile we started building the Nehorah rural center. The idea of a rural center was entirely our own, being aimed at rectifying what appeared to us as mistakes deriving from rural planning in the early 'fifties. An attempt had then been made to mix and integrate without distinction communities, tribes and clans. There were villages filled in one go by immigrants from three or four countries. It was soon discovered that this method would not work. There was friction between the new settlers at every step. Iraqi Jews could not pray in the same synagogues with Hungarian Jews; the cooperative stores in the mixed villages could not supply the conflicting requirements of the inhabitants. Some wanted rice and spices, the others demanded potatoes and fish. Some spoke Yiddish, Hungarian or Romanian, while others spoke Baghdadi, Persian or North-African dialects of Arabic.

We thought to create in Lakhish a new settlement element, which would serve as a more flexible instrument for Integration of the Exiles. That was how we came upon the idea of the rural center. We planned the villages of Lakhish in units of five or six, with an additional settlement in the center of each unit. The extra settlement was to provide utilities for the villages, and would contain a joint grade school, a community center, a sports hall, a bank, a large cooperative store, a clinic, a tractor station, and eventually workshops, such as a smithy, a locksmith, and so on. In every village we would settle members of one community, and every village would have certain minimum elementary utilities: a kindergarten, a synagogue, a small cooperative store and the like; while the more complex services would be provided at the rural center.

Thus, we thought, would be created an area of contact, without strain or pressure. It also seemed to us that six villages with a population of a few thousands could maintain utilities at a higher standard than a single one alone.

Nehorah was to be a first experiment. In Lakhish we also proposed to implement the new type of settlement house; this was also one of our planning crazes.

Until Lakhish, very small houses with an area of 24 or 28 square meters were built in immigrant settlements in Israel — partly for lack of funds and partly on principle. And what was the principle? The veterans and departmental directors said at the time that it was not advisable for the new settler to start off immediately with a large house. Better he should receive a one-room or one-and-a-half-room house, and add the remaining rooms when he is better off. Did not Deganya start from small mud huts at *Um Junni?* And what did Nahalal's first settlers have? Tents and little huts.

One small detail only was overlooked : when the pioneers of Gedera, Deganya and Nahalal settled the land, they were young people unburdened by families and moved by an enthusiastic idealism; whereas our settlers were encumbered by large families with many children, topped by the elderly members of the family — among them the invalids, sick, blind and lame. Such a large family had to make do with the confined space of the miserable blokon while in the neighboring town more spacious quarters were to be found. The narrow blokons drove away quite a few settlers, and we decided that in Lakhish we would introduce a basic change in the dwellings.

We planned a two-and-a-half-room house, with an area

of about fifty square meters. Even this was a modest space for a large family, but it more or less doubled the area of the blokon. Our architects tried to design as nice a house as possible. We had a very tough struggle with the Directorate and Treasury in Jerusalem until we received a budget for building the new houses; but even in Jerusalem they realized that the blokon method could not continue.

Twenty-five houses in the new design were built at Nehorah, and we also constructed a school worthy of the name. The Gubers were the first to move to Nehorah. They helped us in assembling a team of teachers and educators, a doctor, nurses, tractor operators and other public workers. As expected, the Guber home very soon became the hub of life and activity in the place. Rivka immediately proceeded to grow flowers around the house, as well as shade-giving trees. Mordechai set about coordinating representatives from the new Lakhish settlements, for the purpose of establishing an elected council. Thus was set up the first Lakhish Regional Council, headed by Mordechai.

Rivka devoted herself at first to the regional school. She was a perfectionist by nature, and wanted quick results. She insisted on a suitable school uniform, and that the schoolhouse be as clean and polished as her own home. But her concepts and preferences gave rise to differences of opinion between her and some of the new young male and female teachers. Upon leaving the school, Rivka took upon herself the establishment of a library for the village children and adults, in this project, too, investing untiring effort.

Rivka and Mordechai stayed on in Lakhish long after the Lakhish team — directors and young volunteer instructors — had left the district, some returning to their kibbutz home or *moshav*, others moving to other new tasks. The Gubers remained there until 1968, spending thirteen years in Nehorah. They tended the growing pains of the young settlements, and together with them went through their difficult childhood, their problematic growth. They spared not the rod from their foster children, and the latter at times repaid with a measure of bitterness and anger.

And as is usually the case with parents who watch their children growing into adulthood and living in a world of their own, a different and strange world — the Gubers drew the necessary conclusions, retiring following the Lakhish settlements' *bar mitzvah* celebrations. Although their home was at Nehorah, they realized that their very presence there would be uncomfortable to whoever took over from them at the head of the regional council.

The place where Rivka and Mordechai chose to build their home was Kfar Ahim ("Village of the Brothers"), named after their two fallen sons.

Kfar Ahim, near Kiryat-Malachi, had been established by survivors of the European Holocaust, who arrived in the country via the immigrant ships after the Second World War. Among them were even some couples I myself had transported on the immigrant ship *Haim Arlosoroff* — my *Ulua*.

Over the years the settlers had done well, becoming excellent farmers, their livelihood being based mainly on dairy farming. Hearing that the Gubers were about to leave Lakhish, they extended an invitation to them and provided them with a small house in the center of the village.

Again the Gubers planted shade trees around their new home, and once again the flowers bloom in Rivka's garden. Rivka's legs are weary nowadays and Mordechai's blue eyes are dim; but their wonderful spirit is yet with them.

Of late I traveled with them to visit the Georgian Jews who had recently settled at Kiryat-Malachi. Here was another in our multicolored spectrum of tribes, I thought: let the Gubers observe that the work goes on, and that Jews are immigrating and being absorbed in this land. On the Memorial Day for the fallen of the Israel Defense Forces, on the eve of Independence Day 1969, I proceeded with Rivka and Mordechai from their little home at Kfar Ahim to the military cemetery at Kfar Warburg, where their two sons lie buried. They stood leaning on me, two small and elderly people. We stood there in silence for a few minutes.

Rivka then said : "Come, let us join the families of the fallen of the Six Day War, in the cemetery's new section. Maybe we can assuage some of their great pain."

CHAPTER XXII

# THE OLD-TIMERS

Among the volunteers of the *moshav* movement's second generation, who answered Ben-Gurion's call to go out as instructors in the immigrant *moshavim*, were also some of the founding generation, the wonderful pioneers of sixty years of age and more, with over thirty years of back-breaking work behind them.

They heard the call and reported to us — young people in charge of planning and operations. Lost among the hundreds of young male and female volunteers, they yet retained their ineradicable uniqueness.

Looking back across the space of time, they seem to me — with all their apparent anomaly — to be representative of an extinct race; a generation in which there was a fascinating blend of the ways of the 36 righteous men and of the "narodniks," a combination of Rabbi Akiva and of Tolstoy.

I wish to tell the story of two of them:

## RUDI

Rudi was an old friend, who had immersed himself in his farming activity after dozens of years of service in the defense of his country and people. He was then at his kibbutz, Gan-Shmuel, working in its orchard and plantation.

Calling for him, I felt sure that much spiritual and physical strength was still latent in him; or perhaps it was that I simply longed for him.

One day I turned up unannounced at his hermit-like room at the green kibbutz. Rudi was not yet back from the plantation, and I sat down on the doorstep to await him. The room had not changed since I had last been in it : bed made with soldier-like precision, a small table, a chair or two, a bookshelf, a portrait of Orde Wingate on the wall. The owner's exemplary neatness ruled the room; it was as if every article declared : this is my place from which I shall not budge, neither to the right, nor to the left.

And then came Rudi, back from the plantation. The black rubber boots and blue-gray fatigues only served to emphasize his erect form and lean muscular figure. On seeing me, his face lit up, and we shook hands.

My acquaintance with Rudi was then of some twenty years' standing. His fame first reached me when I was still in my apprenticeship in the *Haganah*, serving as a bicycle runner in the 1936 riots. Even then Rudi stood out as an unusual figure among the *Haganah*'s senior commanders. He did not belong to the "*kaffia*-and-dagger" Russian-Caucasian Circassian school of the *Hashomer* organization; nor was he one of the high school boys, that group of first graduates of Tel Aviv's "Herzliya" High School from which sprang the famous family link of Hoz-Golomb-Sharett-Avigur, which also included David Hacohen; nor was he of the "Self-Defense" organization set up in Eastern Europe in the old days, the members of which since immigrated to this country.

To us Rudi was the personification of the professional "Prussian" school. It made no difference that he was neither from Prussia nor from Germany, but from Vienna; to us he was the "German officer" : very thin and muscular.

Rudi was only slightly taller than average, but his erect

gait made him seem tall. He had very blue eyes, gray-blond crew-cut hair, his yellow mustache was very thin and carefully tended. Perpetually in khaki, always very simply and cleanly dressed, Rudi was "anti-slovenliness" and "anti-disorder" personified, in short: a non-Jewish type. So different was he from the other *Haganah* commanders and so little did he talk of his past, that we boys gave him the halo of a legend : the story went that he had been an officer in Franz Josef's Imperial Austrian Army in the First World War; that he had been captured by the Russians and escaped from Siberia, and on his escape route had fallen in with Zionist *halutzim,* was attracted by them and arrived in Palestine with the first lot of *halutzim* of the Third *Aliyah.* This was further embellished : rumor had it that Rudi was not entirely of Jewish stock, and that the blood of German nobility flowed in his veins. And who else but Trumpeldor had influenced him to throw in his lot with Zionism, and contribute of his military experience to the establishment of a Jewish army !

Subsequently, when I became closely acquainted with Rudi, I found that our imagined façade of the "Prussian officer of noble lineage" hid a wonderful person; kind-hearted and sensitive to tenderness, a boundless friend, a fighter of the battles of the just and the weak.

Rudi was one of the commanders of the Field Force in Jerusalem, when I commenced my studies in biology and agriculture at the Hebrew University on Mount Scopus in 1939. I served then as platoon commander in the Field Force, my men being immigrant students mainly from Hungary and Italy.

He was my immediate superior. Together with other field service platoons, we would set out on expeditions through the Jerusalem and Judean hills, going down to *Ein Fascha* and along *Wadi Kelt* toward Jericho, hike to the monastery of *Mar Saba*, studying the terrain well.

Different times were upon us at the beginning of the 'forties, when we joined the various units of the British Forces by the hundreds and thousands. Many of the volunteers were *Haganah* men, who answered the call of the National Institutions to join the Forces.

Rudi, veteran soldier and officer, very much wanted to be among the first volunteers for the British Army; but he bowed to the orders of Eliahu Golomb, Commander of the *Haganah*, to remain behind and serve as chief liaison officer between the *Haganah* headquarters and the members of the *Haganah* in the British Forces.

Throughout the war years, Rudi was a major address for us. He moved from one unit to the next, in Palestine and abroad. He would turn up in the guise of a representative of the Soldiers' Welfare Committee, whereas in fact he coordinated arms acquisitions, communicated orders and received reports.

The war over, Rudi returned to his kibbutz, Gan-Shmuel; very soon, though, he was again called up for *Haganah* work. A difficult period faced the country and its isolated settlements, with the approaching decision on the establishment of the Jewish State. He was appointed liaison officer to the Etzion Bloc settlements, and on the eve of the Declaration of Independence — commander of the Dead Sea region.

Once the War of Independence was over, Rudi returned to Gan-Shmuel.

Came mass immigration. Large *ma'abarot* were set up in the Shomron, and Rudi volunteered to work in them as instructor, attendant, overseer, or director.

Next thing we knew, Rudi had gone down to the Negev. Someone had decided that there was no better place for establishing a settlement than next to the Large Crater. Brown and miserable-looking huts were speedily put up on a hill, in the middle of nowhere, and the place was settled by Jewish immigrants from Romania. Rudi was their sole support in the life of poverty, isolation and helplessness they led. That was the beginning of Kfar Yeruham.

Once again Rudi returned to home and kibbutz. He was no longer young, and his travels were seemingly over.

Sitting in Rudi's room, I told him of the Lakhish dream. I asked him to join the team, and he did not hesitate for one single minute.

Renting a small room in Ashkelon, he started to work with us. During the planning stages he would go out to the territory every morning without the land and soil surveyors. Rudi had a knack for smelling out cultivable soil. He would survey many miles on foot, returning to his room at eventide. Every once in a while he would come to our tiny apartment and play with baby Ophra and little Zvika. Tanya would make him the "German" coffee he was so fond of, and occasionally he would agree to join us for supper, barely touching the food. Tanya very soon came to grow fond of this self-effacing and quiet person.

When the first settlers arrived in the Region, together with the first farming instructors, Rudi became the instructors' instructor. He always found the time to be in the field alongside those struggling with their first furrow, their first sowing, their first fertilizing, and their first planting.

Rudi concentrated his entire energy and attention on the first plantings. There were as yet no areas or water for fruit trees but he undertook the planting of young olive trees along the streets in the new settlements, so that they might eventually provide shade and ornament.

Together with the new settlers, Rudi planted the young saplings he had brought. He did his utmost to protect them, but did not always succeed. The settlers did not grasp the nature of these strange plantings, which did not belong to them personally, the fruits of which they would be able to eat only after many years. "Shade" and "ornament" — who had time to think of these, when there was not even whole-wheat bread in the house and a decent livelihood was hardly available? Those who really took to the olive saplings were the goats, which practically every settler had bought to provide his home with some milk and cheese. To the goats, the olive branches and their juicy leaves were a real delicacy. They were not diet watchers, these goats, and they chewed the saplings right down to trunk and roots.

Rudi and his instructors tried to protect the plantings through friendly persuasion; by employing the threat of sanctions; and principally through the use of barbed wire fencing.

When he left for home after months of toil in Lakhish, he would get satisfaction from the knowledge that his labor had not been in vain. Nowadays, if he wishes, Rudi can go for a stroll through the fields of Lakhish. There he sees thousands of acres of citrus and deciduous fruit trees that he helped plant, now bearing fruit.

The years passed, and I was called upon to set up the Arad Region, in the Judean Desert above the shores of the Dead Sea. Again I traveled to Gan-Shmuel, and waited in Rudi's room for him to return from the field. Again I showed him a schematic chart of what we would some day set up in Arad. A new town, new industry, a new chemical plant, new landscape and new tourism.

Rudi looked at me with his deep blue eyes, and said : "No, this time no. I've no doubt of the importance of it all, but not only am I not young any more — this field of chemical industry is not for me. I'm sure you'll find a team of engineers with whom you can work well."

## The Old-Timers

### "ISAAC"

There are people who are known only by their first name, the surname simply will not stick. And there are people who are never known other than by surname.

"Isaac" was the name which stuck to Moshe Isaacovitch.

Isaac, too, was no longer a young man when he joined us — just like Rudi, the Gubers, Bezalel, Ahavah and the others. He also had many deeds of pioneering and volunteer service to his credit.

He was one of the founders of Kibbutz Ramat Rahel, on the outskirts of Jerusalem. Toward the end of the 'thirties, he joined the youngsters at Kibbutz Beit Ha'arava on the Dead Sea's shore — in the backbreaking task of washing the salt out of the dead land, transforming it into life-producing soil. After the War of Independence, when Beit Ha'arava had to be abandoned, Isaac returned to Ramat Rahel. It had meanwhile been almost completely wiped out in the attacks and shelling of the Jordanian Legion and the Egyptian forces.

While to the South, Negbah's brave stand had halted the Egyptians' progress, the fall of the Etzion Bloc on the eve of Independence had left the way clear for their advance northward along the central ridge; Ramat Rahel had, in fact, changed hands a few times during the War of Independence but it stood fast at tremendous cost in life and saved Jerusalem.

A handful of Ramat Rahel veterans set out to rebuild their kibbutz and Isaac was their spokesman. During my period of work with Eshkol, I would see him practically every week, chasing after Finance Department staff, directors and engineers — urging, interceding, begging and threatening. Isaac had connections and access to all the top people, from Eshkol

161

and Harzfeld down to the veteran clerks, who had known him in the early days. Thus Isaac secured for Ramat Rahel what no one else could have.

I lost sight of him from that time and until the advent of the Lakhish Region, when he turned up to offer his services. I did not delve deeply into what had happened to him at Ramat Rahel. Gradually, I learned that he had reached a crisis, and had uprooted himself, his wife and sons from the place he had twice labored to found.

I consulted Levi and Benny as to what to do with Isaac. He was still in full strength, and greatly experienced in management, as well as in work in unconventional settlement projects — to whit, Beit Ha'arava. We decided to place him in charge of the economic aspect of the *Nahal* outposts to be established in the Region.

At least three such outposts were planned for the eastern border of Lakhish: one next to Tel-Lakhish; the second, to be known as Amatziyah, right on the Jordanian border, near the abandoned Arab village of *Douema;* and the third, Nehoshah, also on the border, on the road which once linked Beit Govrin with Hebron.

Isaac was put in charge of these three outposts. It was his job to attend to all their financial requirements.

The three border outposts took root one after the other in the summer of 1955. Because of security conditions, the young *Nahal* settlers existed in a "Wild West" atmosphere; in fact, the Army gave the district the nickname of "Texas." The place was infamous for being wide open to infiltrators from across the border. The many caves in the area provided shelter for the infiltrators and terrorists, who dared to snipe even in daylight. The *Nahal* soon had its share of fallen soldiers, and it was not long before the first victims were claimed from among the young settlers in Lakhish.

Isaac chose as his residence an old and dilapidated house

which stood in an abandoned orchard on the outskirts of Ashkelon. Already then he had ideas for building a large family ranch among the groves of Ashkelon. There was no electricity or water in the house. Isaac was insistent however — for he was a stubborn man — and brought his good and dutiful wife and his sons to the place.

Every morning at dawn, Isaac would set out for the *Nahal* settlements, remaining there all day and sometimes staying overnight. He would dash about on matters concerning their financial affairs and the agricultural work, just like a mother hen with her brood. His extremism sometimes bothered the *Nahal* boys and the Region's staff; he actually accosted the personnel of the Region's different departments. The team members would come to me with complaints : "He needs everything, this Isaac of yours; and everything fast, and everything immediately, and he never has time." Isaac nagged the tractor operators, following them and their plows, and insisting on greater depth and more acres plowed. He saw to seeds and fertilizers. He nagged Rafi Gurevitch and his men for small pumps to work the water cisterns, as well as the old wells found in the vicinity. He also watched over the *Nahal* settlements' "cash box," and was a miser of the first degree. "Not one cent must be wasted," he would lecture the boys, "lest there be deficits." The money was to be reserved for constructive purposes.

Isaac established the first livestock farms at the outposts : flocks of sheep at Lakhish and Nehoshah, and a cattle herd at Amatziyah. Experiments in cattle herding were then being conducted in Israel, and the Lakhish Region was chosen as one of the locations. A herd of "St. Gertrude" breed was introduced to the Region, strange-looking cows and bulls from the United States meandering through the fields of Lakhish like creatures from another planet.

The first *Nahal* settlement to start standing on its own

feet and to be released from Isaac's authority was Tel-Lakhish, which was inhabited by a group of second- and third-generation *moshav* members, mostly in their twenties. I got to know them in the first days of my work in the Negev, at which time they were settled temporarily at Sde Boker. David Ben-Gurion had discovered them, and took to them from the first glance. One of them had been the "Old Man's" instructor in sheep herding.

These young *moshav* people now decided to transform their outpost into a permanent settlement. We consulted jointly on all planning details, starting with houses and ending with seed rotation. Within a few months, the army-style shacks were replaced by permanent structures. After much toil and trouble, Tel-Lakhish was to become one of the Region's pleasantest and best established settlements.

Amatziyah and Nehoshah were more problematic settlements, due to the instability of their population. Settlers came and went, and Isaac's hands were full of work.

But Isaac also had ideas for a place of his own in the Lakhish Region. He pleaded with us, almost desperately, until we gave in and allocated him a "parcel" of land of about one hundred acres in the western part of the Region, near Ashkelon. We helped him in securing loans and in setting up a farm based on groves and truck gardening, with five houses for him and his little tribe — five families in all.

Isaac started afresh. An elderly man who had experienced trials and troubles and now wished to devote himself to his home and family.

CHAPTER XXIII

# DEBORAH DAYAN AND HER SONS

I first heard about Deborah Dayan, Moshe's mother, from
her brother, Yehoshua Zatulovsky, during the Second World
War. He was then a lieutenant in 745 Company of the Royal
Engineers, and my platoon commander.

Yehoshua himself was a real character. When he volunteered for the British Army, he was already in his late fifties.
A Russian Jew through and through, his Hebrew had a very
pronounced Slavic accent. He had come to this country as
a pioneer with the Third *Aliyah*. Where had he not worked
and in what hard labor had he not engaged? Over the years
he had become one of the leading construction foremen of
*Solel Boneh*. When the *Solel Boneh* men were conscripted to
745 Engineering Company, Yehoshua — who was well above
army age — volunteered, and in accordance with an arrangement with the Army authorities, as well as by virtue of his
seniority, was automatically made an officer.

I was over thirty years his junior, and a young corporal
in his platoon. Being a veteran of some two years' Army
service, I became the platoon's armed drill and light weapons
instructor, and I instructed Yehoshua, too, in military ways
and in the use of weapons.

He was of medium height, tough and broad-shouldered,
and his face was lined and bronzed from years of work on

building sites. Yehoshua loved a friendly chat over a drink
with his soldiers after a day's work. Thus, in N.A.A.F.I. clubs
in British Army camps along the Suez Canal or in cafés in
Suez, Ismailia and Port Said, I heard tales from Yehoshua
about the early days in the Ukrainian village where he was
born and about the beginnings of the Zionist Movement, as
well as a multitude of stories of working life in Palestine in
post-World War One years.

It was also then that he spoke of his sister Deborah, who
had preceded him to Palestine, immigrating in 1913, and who,
with her husband Shmuel, had been among the founders of
Deganya, and later of Nahalal in the Valley of Israel.

Some years later I became closely acquainted with Zorik,
younger son of Deborah and Shmuel.

Italy, 1946. Zohar (Zorik) Dayan arrived at the Italian
branch of Mossad, the *Haganah*'s arm for organizing illegal
immigration. He reported to Yehuda Arazi, the commander
in Italy. Zorik had volunteered, like many of us, after service
in the British Army's Jewish units. Being very young, he had
not seen much service with the Army; the war over, Zorik
plunged into the Yishuv's fight for the rights of Jewish immigration.

Those summer months I was appointed, by virtue of seniority, to command a sailing camp located in a château
we had rented. The property of a rich Italian, the château
was one of the most beautiful of its kind and was situated
in the village of Bogliasco, south of Genoa in the Italian
Riviera.

I sought a team with which I could work on fitting out
the château, and to prepare the little inlet nearby for the
influx of immigrants and for boats. Zorik was detailed for
this task, and we worked together for a few months at Bogliasco.

Zorik was then a handsome young man of about twenty.

His athletic build, plus his wavy chestnut hair, made him seem taller than his medium height. He was an excellent driver, very fast, though possibly not as careful as one might wish; he had full mastery of the wheel of his jeep and knew its innards well.

On the face of it, Zorik was a typical *moshav sabra;* tough, brave, open-hearted, untalkative, stubborn, a born field man and a born mechanic. Behind this appearance, though, there was a different person — a gentle and talented poet, acutely sensitive to people and landscape.

Zorik had a sweetheart back home — Mimi. Our acquaintance with her at that time was confined to the letters Zorik wrote her and the poems he dedicated to her. Until the advent of the first immigrants, we were three Israelis alone in the secluded château, and we had no secrets from each other.

A few months after the château had been transformed into a sailing camp, I was transferred to a different post and never saw Zorik again.

Upon my return home in 1947, I heard that he had married Mimi, that they were expecting a child, and that they had made their home in Nahalal. It was not long, however, before I, too, was a happily married man, and 1948 was upon us, and none of us managed to see our friends.

Zorik fell in the fighting at Ramat-Yohanan.

I got to know Mimi later, when she was already mother of a baby boy called Uzzi. Some of Zorik's friends, myself included, tried to draw Mimi nearer to us; we were then officers in the fledgling Navy.

One year after the War of Independence, Mimi married Moshe Rabinowitz, one of our finest men in illegal immigration work and later in the Navy. Mimi and Moshe built their home at Moshav Hayogev, which had been established close to Afula in the Valley of Jezreel. Moshe tended the farm, while Mimi with much talent set up a small workshop for fine metalwork and gems.

It was at their home that I made the acquaintance of Uzzi's grandmother, Deborah Dayan.

Deborah was then slightly over sixty, a short and very thin woman. Her lovely black hair and lively eyes were a reminder of her youthful beauty, of which her brother Yehoshua spoke.

I was then working with Levi Eshkol, and Deborah had heard of me from her husband Shmuel, who was very active in the *moshav* movement. Zorik possibly, and Mimi and her Moshe certainly, told her about me. We soon became friends, Deborah also having taken up organization and instruction activity in the immigrant *moshavim*.

We discussed many things: the situation in the new villages, how to improve, how to correct, how to ameliorate. Deborah was very purposeful in her conversation, straight and to the point.

It took me some time to plumb the kind-heartedness and depth of feeling beneath the shell of matter-of-factness — the outcome of years of hard toil in raising family, home and village, and deriving from a tough and strict daily regimen. She had above all the willpower to fulfill this difficult way of life.

It was then that I also understood Zorik's complex nature, and in the course of time it provided me with a key to understanding — albeit partially — the personality of the eldest son, Moshe.

I made Moshe Dayan's acquaintance only at random and at chance General Staff meetings during the War of Independence. Our paths moved in parallel courses: Moshe joined the *Haganah's* special units, while I served in the British Forces; he was indubitably an infantryman, whereas I landed in the Navy; and later, when I took up settlement work,

Moshe continued with a military career, rising fast up the rungs of command.

Our first real meeting was before the setting up of Lakhish, when he was already Chief-of-Staff and had become active in the public movement initiated by Ben-Gurion: the call on *moshav* and kibbutz members to volunteer for instruction among the immigrants. He had tremendous influence on his own generation and on the younger people in the veteran *moshavim*, and employed it to the full.

The volunteer movement had all the ingredients and shades of an ideological movement; it had leaders, committees, social problems, lengthy debates on its course and, of course, many organizational problems.

Moshe would attend their conferences as one of them, delving with them into the thick of things. Often I saw him arriving at these meetings in the Chief-of-Staff's car; he would remove the military insignia from his shoulders, enter, and sit among his comrades dressed in simple khaki, sans rank and decorations.

I was an instructor at Nevatim when the movement had its beginnings, and its leaders adopted me as one of their own. It was then that I had my first meeting with him at close quarters.

When the idea of settlement in Lakhish first took shape, I obtained his help in a rather unconventional manner in everything pertaining to the Israel Defense Forces, and the call-up of volunteer instructors.

Deborah joined us in the early days of Lakhish. Her health was gradually being undermined. She seemed to shrink and her face was a chart of wrinkles deriving from years of toil, too much grief and too little happiness.

I would travel with her through the immigrant *moshavim*. Her principal business and concern was instruction of the women, and she urged me incessantly not to relax nor neglect attending to the women's instruction, many of whom could neither read nor write, most of them burdened with children and household chores.

Deborah was always trying to secure a bit more money especially for the women — to get rooms for instruction in housekeeping and home economics, a mother-and-child clinic, school, dining rooms and so forth.

The trips along the dirt tracks tired her out, and she was pleased to accept my offers to visit our home in Ashkelon any evening she liked and to rest a while before the next working day. Thus Deborah also became a family friend.

Before many months, Deborah Dayan fell victim to the illness from which she never recovered. Her strength gradually ebbed, and her visits to the Region became less frequent.

On one of her last visits, we had a long talk about what it was then the custom to refer to as the "generation of the sons" — that of Moshe and Zorik, her sons.

Deborah said: "Moshe will not remain forever in the Army, nor you in Lakhish. In a few years' time, you and your likes will start bearing the heavy burden, the extremely heavy burden, which the Jewish People imposed on my generation and under which we are already sinking."

I told her that I proposed to terminate my work in Lakhish within half-a-year, transferring management of the Region to Levi Argov, and taking up my studies again at the Hebrew University of Jerusalem. Deborah congratulated me on my decision, adding that she hoped Moshe would also manage to acquire a higher education.

A short while after this conversation, Deborah was admitted to the hospital and about a month later she passed away.

During the first two years of Lakhish, Moshe Dayan visited

the Region and its settlements a number of times. Naturally, his visits at first were mainly to the *Nahal* outposts on the Hebron border: Lakhish, Amatziyah and Nehoshah. The entire district was troubled, constantly invaded by infiltrators, and the Army had its hands full.

I would join him occasionally, and the beginning of a friendship was forged. I told him of the problems of the immigrants' absorption in the *moshavim*, and of the conflicts between them and their instructors. Many of the latter were, of course, comrades and pupils of his, and occasionally we also paid them surprise visits in the villages.

I told him of the long evening and night sessions the instructors and I spent with family-clan heads in the villages, and what a committee meeting in an immigrant *moshav* appeared and sounded like. He requested me to take him along to one of the sessions; having fixed a date, we set out in two jeeps for Moshav Sdot-Micha.

A few dozen Moroccan immigrants and their instructors awaited us in the tiny secretariat hut. On the agenda: election of their village committee. The settlement was only a few weeks old and was one of the most distant and isolated in the Region's north-eastern hilly district not far from the Beit Govrin-Beit Shemesh road.

We took our seats with the immigrants pushing and crowding around.

There was no dearth of candidates for committee chairman. The ramified Levi clan proposed one of its own members. This family had a near majority among the village inhabitants. However, the other candidate — of the smaller Cohen family, with less chance of being elected — had succeeded in forming a coalition with the Ben-Haroush family, and together they planned that evening to topple Levi and his family, and elect their own man as committee chairman.

We opened the session, and the chief instructor gave the

171

floor to the settlers. Every one of the speakers, and they were many, started off with a long list of complaints about the situation in the village, and then went on to praise his own candidate. The entire affair was conducted partly in basic Hebrew, but mainly in Moroccan-Arabic which was translated into Hebrew for our benefit.

Moshe Dayan listened attentively to the first speakers, glancing at the settlers with curiosity. After half-an-hour it was clear that he had caught on to the principle, and that it sufficed him as a lesson in "democracy in action." Rising, he exchanged a few greetings, and then drove off in his jeep into the night.

The next time we met, he asked me: "How did that evening at Sdot-Micha end?"

I told him about the dramatic night of the election of the committee chairman and members.

"And what time was the session over?" he asked.

"At about 1 a.m.," I answered.

"Where do you find the patience to sit with them, day after day, and night after night?" he asked.

Subsequently, when Moshe became Minister of Agriculture, we were trying to get hold of Ephraim Shiloh — a member of Kibbutz Tirat-Zvi and one of the country's foremost agricultural experts — for the rehabilitation and development project of Iran's Kazvin region, which had been destroyed in an earthquake. I had been appointed to head the project. As was its custom, the kibbutz was obdurate and did not tend to accede to my request to release Ephraim for outside work. I told Moshe about it.

"Come, let's go to Tirat-Zvi and talk to them," he said.

We advised the kibbutz of our coming, and arrived toward evening to attend a general meeting there, the agenda being: "Conscription of comrade Ephraim Shiloh." The dining hall was full of members, with members' parents and children also turning up for the show. It was also an opportunity for the female members, while attending the deliberations, to finish knitting scarves and vests.

I commenced by speaking of the importance of the work in Iran, and Moshe backed me up. A discussion developed, with the usual two schools of thought forming — the pros and the cons. The discussion went on for quite a while, Moshe and I participating and patiently reiterating our arguments. The decision was finally reached after midnight: the kibbutz decided to release Ephraim for outside work, placing him at the disposal of the Iranian project.

Returning after midnight from the Beisan Valley to Tel Aviv, I reminded Moshe of those nights in the immigrant *moshavim* and commenting, "The path from 'democracy in action' in a kibbutz to the election of a committee in an immigrant *moshav* is a long one, but in the final count it's the same path. We shall yet live to see the time when we have to call on the immigrant *moshavim* for outstanding experts for agricultural planning and instruction in Israel and throughout the world."

And, indeed, such times arrived sooner that we expected.

## CHAPTER XXIV

# HAMASHBIR'S LEAN YEARS

The regime of austerity and rationing was still in full swing when we opened the cooperative stores in the immigrant *moshavim*.

Israel's pre-1948 farm economy was unable to half fill the food-basket of the burgeoning population, as the new immigrants poured in.

Vegetables, fruit, milk and dairy products were all rationed. The immigrants in their new *moshavim* were still by way of being consumers only, their fields — eventually due to yield of the best — still only grew weeds. Other foodstuffs such as rice, sugar, oil, wheat, fish and meat had to be imported and paid for in hard currency. The State Treasury was bone-dry, Finance Minister Eliezer Kaplan having long since used up or mortgaged the few reserves of Sterling which survived the war. Oranges were the sole export item for which there was a market and which brought in some cash, but the yields were long since mortgaged. In order to meet the cost of a ship-load of wheat, Kaplan had to tap all his own and his colleagues' financial wizardry. Urgent phone calls and cables would go out from Jerusalem to New York and Geneva, in order to raise $ 10,000 to pay for a load of sugar or rice en route to Israel. All the while suppliers were threatening to instruct the captain to deviate from his course and not to unload the food at Haifa.

Little wonder, therefore, that the Government's "Joseph the Purveyor" — Food and Supply Minister Dov Joseph — imposed on the country a regime of the "seven lean years." All food items were strictly rationed. Spotted, speckled and dappled ration cards were issued to all families, and grocers became bookkeepers on a small scale, wielding scissors to cut out multicolored coupons. Inspectors descended on road and settlement, investigating every vehicle lest it be smuggling food products.

It was at this time that the cooperative stores opened in the immigrant *moshavim*.

The cooperative store was established on the ideological basis of consumer participation in profit; as a perfect cooperative institution, it was one of the components of a modern system which the labor movement applied in the collective workers' settlements. In the immigrant *moshavim*, it was an out and out artificial institution. In reality it was a small grocery store occupying a shack in the village center, receiving its goods from the *Hamashbir La'Oleh* chain, which had been established as a subsidiary company of *Hamashbir*. The consumers, the fifty or sixty immigrant families, were very poor, with near-zero purchasing power. With the decision to establish such a cooperative store in every *moshav*, it was obvious even to anyone who had not attended a business administration school that the entire venture was doomed, and *Hamashbir* was constantly demanding subsidies from the Jewish Agency.

The new settlers were not aware of this. All they saw was a store with perpetually half-empty shelves, the other half containing what to them were tins of dubious preserves, not having been acquainted with them in their countries of origin.

What appeared to them even stranger and more annoying were the store clerks. The immigrants soon realized that they were dealing not with ordinary grocers, traders with whom one could haggle, but with hired clerks whose attitude was "take it or leave it!"

The cooperative store raised the ire of immigrants from the Communist countries, since it reminded them of the *Univermags*, those infamous Soviet "supermarkets" in *kolkhoz* and *sovkhoz*. And it was an even stranger phenomenon to those who came from Islamic countries. Instead of the grocer at the counter in the colorful and noisy bazaar, they were faced with an anonymous clerk in an anonymous hut, who handed out such peculiar food items as yoghurt, frozen fish fillet and the like.

A very problematic subject was that of the fare the cooperative stores were called upon to supply people from a hundred and one different countries of origin. The Poles wished to prepare gefilte fish, Moroccans wanted the ingredients required for making couscous, Romanians hankered for *mammeliga,* Indians could not exist without curried rice, Yemenites wanted the components of *falafel,* Russians thirsted for borscht, Hungarians hungered for goulash with paprika, Persians wanted to prepare their favorite kebab, and so on. The store clerk's stock reply to all of them was: yoghurt and frozen fish fillet.

We saw to it that at least one staple food, bread, should arrive on time and in more or less sufficient quantity. Admittedly, man does not subsist on bread alone, but if there is flour there is no hunger, and at that stage of village life this was the principal concern.

On the whole, the bread was fairly fresh when it reached the Lakhish settlers. Brought from the *Hamashbir* bakery at Ashkelon, it was distributed to the cooperative stores by a pickup truck which made the rounds of the villages.

One day — it was the eve of the first Feast of Weeks in the Region, and we were preparing to usher in the Feast in our little home in Ashkelon's *Afridar* quarter — a number of instructors turned up suddenly, announcing excitedly: "No bread has arrived today at Nogah, Shahar, Otzem, Sde-David, Eitan or Noam. The settlers are going out of their minds. What's to be done?"

"What happened? Why did the bread not arrive? Have you checked at the bakery? Did you try to buy elsewhere?" I asked.

"We waited for the pickup truck all day, thinking it had had a temporary mishap," the instructors replied. "In the afternoon we drove to the bakery at Ashkelon, and found it closed. We called on the manager, and he told us that there was a technical fault in the oven, and no bread could be baked."

I wanted to ask why the bakers had not informed them of this earlier, when we might have got bread from another bakery, possibly at Rehovoth or Rishon-L'Zion. By then it was too late.

We consulted as to what was to be done, when all of a sudden I had an idea. Stepping into the kitchen, I asked Tanya how many loaves of bread we had.

"Four loaves," Tanya replied.

"Good! I'm taking three and leaving you one. *Shalom,* I'll be back in a few hours."

From my home we drove to Haviv's home. Going through the same procedure, we confiscated four loaves from his wife. We repeated the exercise at the homes of all our personnel living in *Afridar,* and within half-an-hour our two jeeps were loaded with dozens of loaves.

Stepping on the gas, we reached the *moshavim* just as the Feast was being ushered in.

Passing from hut to hut, we uttered a festive greeting to the assembled families and left a visiting card — a loaf or half a loaf in every hut.

It was nearly midnight by the time I got back home. I

177

dialed the home in Tel Aviv of Ya'acov Efter, Managing Director of *Hamashbir*.

"*Shalom*, Ya'acov Efter, a happy holiday to you," I said. "This is Lyova Eliav speaking, Director of the Lakhish Region."

"*Shalom*, Lyova, and a happy holiday to you, too," Efter replied, completely surprised. I had the impression I had awakened him from the sleep of the just.

"Ya'acov, I recall when I was a youngster, studying the history of the country's labor movement, I read the story of how you founded *Hamashbir* on the eve of the First World War."

"Yes?" Efter responded politely, thinking no doubt I had drunk a bit too much, and was thus in a good mood.

"I recall the wonderful story of the sack of flour you bought in Tiberias, carrying it on your back all the way to Deganya and Kinneret, bringing some flour to the hungry and impoverished pioneers."

"Quite right, that's how it was," said Efter, still wondering what I wanted of him at that hour of the night.

"We were told it was then, that day, you decided to establish the institution which would provide the needs of the settlers in this country, and that that was the beginning of *Hamashbir*."

"That, too, is correct," said Ya'acov.

"Well then, listen to me," I said. "The settlers nowadays are the new immigrants in the new *moshavim*, and it is for their sake that both we and *Hamashbir* exist."

"Right again," said Efter, with growing impatience.

"In that case, please help me to carry on the tradition you started of serving them." And here I told him the story of the bakery.

When I had finished, Ya'acov said: "Go to the town's bakery manager, and tell him that I shall be arriving from

Tel Aviv tomorrow morning with expert technicians. I hope that by tomorrow night we will be able to repair the oven and bake bread."

And so it was. The following night, as soon as the Feast was over, they were already baking bread in Ashkelon. Efter was there, issuing orders and advice. The man was old, and for many years had been running a huge concern with a turnover of hundreds of millions of pounds; there and then, though, I sensed, he had suddenly rediscovered the purpose of his life's project.

## CHAPTER XXV

# FEDAYEEN NIGHTS

When, in 1955, we undertook the establishment of the first twenty settlements in Lakhish, with hundreds of huts sprouting on the hills and thousands of people — men, women and children — settling and occupying them, when the big plows started taking to the terrain, followed by the bulldozers, dredgers, ditchers, steamrollers — all the modern monsters, the "Texas" days were over in Lakhish.

The infiltrators encountered new facts that cropped up from time to time in the territory. They were no longer able to move freely across open wastes. On the other hand, they found new scope for theft and robbery among the new settlements. And then, in the summer of 1955, there commenced the attacks of the *fedayeen*, sent by the Egyptians to murder Jewish settlers and travelers within Israel's borders.

Israel Defense Force units started pursuing *fedayeen*. The famous Givati Brigade took up positions in the district.

We were anxious about the immigrant settlers. How would they stand up to this threat, knowing nothing whatever of matters concerning guard duties, defense and security?

A district commander, a young sergeant or officer in the reserves, was appointed to every team of instructors, to be in charge of the new village's security. The other instructors were at his disposal for night watch duties and for liaison with the Army and Police in the vicinity. His second important task was to train the settlers, the younger ones in particular, as fast as possible in the use and maintenance of weapons and in guard duties.

We knew that a night watch of four or five volunteer instructors — even with the addition of a few settlers, having only just learnt to hold arms and never having experienced fire — was insufficient to safeguard a village of a hundred families living in a hundred huts dispersed over a wide area.

There was no electricity in the villages, and we did not even dream of fence-lighting. At nightfall, a great darkness would descend on our little villages. The settlers sensed the tension, and when it became dark would call in the children, shutting and barring doors and windows, locking themselves in behind bars, and not answering any call.

The *fedayeen* started ambushing cars and trucks on the Region's roads. At Moshav Shafir, a hand grenade thrown into a settler's home killed all the occupants.

Our anxiety for the settlers' fate increased. One morning I set out for Metzudat Joab, where the Givati Brigade headquarters was then located. I requested an audience with the Brigade Commander, Lieut.-Col. Yosef Geva, whom I knew well from the time of the War of Independence.

"The Brigade Commander is asleep," the duty officer told me.

"Please wake him up. I've an urgent matter to discuss with him."

"But the Brigade Commander is tired out, following a night of patrol duty. He requested not to be disturbed, unless the matter was extremely urgent."

"The matter is extremely urgent," I said.

After some further discussion, I was allowed to enter the Brigade Commander's room.

Yosef lay there on an army cot, wrapped in a blanket

which covered his face, not letting in the light of day. The instant I touched his hand, very lightly, he opened his eyes. Seeing me, he unwound from bed, fully awake and ready to hear what I had to say.

I told him of the situation in the settlements, dwelling in particular on the people's mood, which was turning into panic. I told him about the people shutting themselves in at night, and of the fear which was gradually spreading throughout the villages.

Yosef heard me out, nodded his head, and stated that the Brigade was doing all it could in pursuing and capturing the *fedayeen*.

Telling him that I did not for one moment doubt the Army's efforts, I added that it was not enough for the Army to be doing things — it was necessary also that the settlers sense it.

"How do you mean, sense it?" Yosef asked. "Do you want us to let the heads of families participate in ambushes and pursuits?"

"No," I said, "the Army's presence must be illustrated to the settlers."

"What do you want me to do, hold parades through the villages? Or have a display of firepower? Or what? Believe me, my soldiers are sinking under their burden."

He thought for a moment, and said: "Perhaps, if you wish, I'll recommend that some *Gadna* members be brought to the villages, to liven things up."

"No, Yosef," I said, "I want something totally different! And I'm aware that what I'm asking for is no simple matter and is quite unusual: I want you to place at my disposal — that is to say, at the disposal of the Region — three or four half-tracks with their personnel and let us go round the villages with them every night. I know there's no great operational value in such rounds; but you cannot possibly imagine their psychological effect, and to what extent it may raise the settlers' morale."

After a long moment's silence, Yosef said: "Do you know what you're asking? I have no more than about a dozen half-tracks in good condition in the Brigade, and their personnel is on operational duty day and night. You're asking me to tie up a third of this force for the morale of your people. With all due respect, you're asking the impossible. And even if I could help you in any way, I'd have to get permission from the G.O.C. Command."

"Yosef," I said, "I can't leave empty-handed; I simply can't. You know what? I'll make do with two half-tracks; and not all night, but half the night. Surely you can incorporate these half-tracks in your current operational program? After all, the distances aren't great; let them do one round here and one round there. And you needn't give me your newest or best half-tracks; the important thing is they should travel, and make noise, and have a projector, and there should be a few soldiers in them with rifles and a large machinegun jutting out. I don't need any more. As for the G.O.C. Command, let's agree in principle, and I'm sure if I tell Meir Amit about it, he'll give you the green light."

"One half-track," said Yosef.

My face lit up.

"When do you want the half-track?"

"What do you mean, when? Tonight, of course!"

Yosef agreed.

Toward evening I returned to Metzudat Joab, dressed in a heavy khaki overall, wrapped in a warm vest, a balaclava helmet on my head. The half-track stood in the courtyard.

"Meet Lieut. Moshe Levi," Yosef said to me, introducing a black-haired, lanky youngster. "He's your half-track's commander, and has been instructed to set out with you to the villages for the entire night. He's in communication with Brigade Headquarters and with other units in the area."

Shaking hands, we climbed aboard, and the half-track moved off.

After a quarter-hour's drive, I requested Moshe to park at the side of the road. I explained to him briefly what was going on in the villages, and the purpose of the strange patrol ahead of him. Moshe grasped the situation immediately. He told me that he himself had been born in Baghdad, and had immigrated to Israel not so many years before. Together we planned the night's trip.

Moshav Otzem was shrouded in darkness when we approached it. The half-track plunged off the road and onto the dirt track leading to the village. A cloud of dust arose behind us and a shaft of light groped ahead; Moshe operated the projector, directing the moving light toward the *moshav* huts.

The engine roared, the driver deliberately increasing the noise. Upon our arrival at the village center, we were greeted by the instructors, headed by Bezalel. Jumping off the half-track, I explained to them the purpose of the exercise. Remounting the vehicle with some of them, we drove up to the settlers' darkened huts.

Stopping with a grinding of brakes outside the hut of Rabbi Moshe, a village elder, we knocked on the door. Once Bezalel had identified himself, Rabbi Moshe unlocked the door, his wife and five children crowding behind him, curious and anxious. We stayed there for about half-an-hour, drinking pungent Moroccan tea; the woman of the house took steaming glasses of tea out to the soldiers in the half-track, who continued to operate the projector, sending shafts of light in all directions.

We chatted with Rabbi Moshe about conditions at Otzem, about work, the cooperative store, the houses being built. We also told him something about the Army and its strength.

From Rabbi Moshe's hut we proceeded to the other end of the *moshav*, where we repeated the exercise at the hut of Rabbi Yitrah, another clan leader and village elder. We toured the crude roads of Otzem with our tracks; we dispersed fear, and the huts shook as we drove by. Parting from the instructors, we left Otzem after an hour and set out for Nogah.

We proceeded to the hut of Rivka and Mordechai Guber. There were tears in their eyes when they saw us. As though she had been expecting us, Rivka produced food and pastry for all of us.

The Gubers accompanied us to Yehuda Sharifi's hut, and from there to *Hakham* Yehezkel's. By then, Lieut. Moshe and his men fully appreciated the importance of what they were achieving by their night patrol through the *moshavim*.

From Nogah we traveled to Shahar. After midnight we proceeded to Sde-David, and from there to Heletz and Tlamim. At 3 a.m. the half-track's noise awoke the inhabitants of Eitan. By the time we got to Uzza, Ahuzan and No'am, dawn was breaking; the skies paled, and there was a reddish glow to the east. Switching off the projector, we returned to Metzudat Joab.

I said to Moshe: "Let the men sleep now. Tonight I shall be sending my assistant, Levi Argov, with you. And, mark you, this is what we shall do every night."

"Until when?" Moshe asked.

"Until when? Until such time as the instructors inform me that the settlers know how to handle weapons, and it becomes apparent that their spirit is steadfast."

The lonely half-track went the round of the settlements for many nights. In October 1956 the Israel Defense Forces broke through to the Gaza Strip and the Sinai Desert. The *fedayeen* bases were destroyed, and the Region was peaceful for another ten years.

## CHAPTER XXVI

# FIRST STEPS

Before Lakhish, it had been the custom to introduce the new immigrant settlers to the difficult burden of running an independent farm, without any intermediate stages. And not just any independent farm, but one within a modern and sophisticated cooperative context, such as the *moshav*. This complex form of settlement had been moulded and crystallized in Israel over two generations, in a slow process of trial and error; the brainchild of such people as Eliezer Yaffe, Shmuel Dayan, Ya'acov Uri and others. The *moshav* of today, with its rules and regulations, has a complicated system of accounts involving credits, loans, interest, financing, supervision and control, purchasing and marketing organization. They are all combined in a regional and national network of such agencies as *Tnuva, Hamashbir,* the Workers' Bank, Settlement Department (budgets), Agricultural Center, *Moshav* Movement, and many more.

Little wonder that the newcomer who settled on the land after the establishment of the State, looked upon this system with a mixture of suspicion, and bewilderment, floundering therein like Jonah inside the Whale.

East European survivors of the Holocaust, behind them years of confinement in concentration camps in Germany and Siberia and of wandering in the steppes of Soviet Central

Asia — these were among the first to be sent to the new *moshavim*. Their minds were immediately beset by peculiar comparisons between the new bodies, movements and institutions surrounding them in their new country, and the *kolkhozes, sovkhozes, sovnorkhozes* and *gosplans* they had experienced in their years of wandering.

The Oriental Jews, too, who went to the *moshavim* in the early 'fifties, were unable to understand the new framework, so strange to their ways of life and concepts. The following is a good illustration.

After arduous work, a person succeeds in growing his first half-ton of tomatoes. The tomatoes have turned out well, and he is ready to load them on the donkey-cart to sell in the market. Hold it, though! This contravenes the rules of co-operation (what a strange word!). He has to wait until the other members of the *moshav* have also harvested their crop, whereupon a truck belonging to *Tnuva* turns up from somewhere, and his tomatoes are taken from him and loaded with all the rest; the truck departs, to the Devil knows where, and he is left with a slip of paper in his hand. A week passes, and he asks the instructor: "Where's the money?" The instructor replies: "Patience." A fortnight goes by, and again he asks, annoyed: "Where's the money?"

"Patience," says the instructor. Two more weeks pass, and finally an account slip arrives from *Tnuva*. The settler goes nearly crazy with anger: There is everything in the slip but money. There is "gross" and "net" and "tare," there is "depreciation" and "administrative expenses," there is "grade A," "grade B" and "grade C," there is "culls," there is "deductions for committee expenses," and any number of other entries — but cash there is none.

A terrible suspicion arises in the settler's mind. There has been a conspiracy against him to rob him of all his possessions, to sell his produce for a song, just so the "directors" can make a decent living. Such frustrations led more than once to the undermining and abandonment of the immigrant *moshavim*.

We wished to get through the stages of adaptation and prevent crises.

Our idea was to create some sort of provisional barrier between the new and inexperienced settler and the harsh economic system of supply and demand, on the one hand, and the *moshav*'s cooperative set-up, on the other.

In accordance with the *moshav* movement and the other settlement groups, we set up development ranches. Instead of immediately dividing the *moshav*'s agricultural square into 80 or 100 farm units, the entire square was to be cultivated as a single large ranch. The settlers would occupy their homes, but would not get their plots; instead, they would be employed on a daily-hire basis as farm laborers on the large ranch, which would be managed by experienced farmers from the veteran *moshavim*. We were aware that these ranches would not yield large profits, certainly not in the first year or two, with our laborers unskilled in modern farming; but this way we would save them initial failures, and prepare them for independence.

The first year was difficult. It was necessary to teach the settlers the ABC of plowing, cultivation and harrowing, the rules of irrigation and of handling pesticides, as well as the proper use of seeds and how to manage an orchard.

The first crops did not excel in quality, but the vegetable-hungry market those days absorbed everything. The settler was not bothered in any way with marketing and selling; he received a daily or a monthly wage, and was free to acquire skill in field work without worry. Spring, summer, fall and winter went by, and yet another spring, summer, fall and winter passed. The cycle of the seasons was absorbed into the new settler's consciousness and way of life, and the cultivated areas were extended and covered with irrigation networks, acre by acre.

The moment arrived when the instructors in the first-born *moshavim*, Nogah and Otzem, began sensing a new wave of tension and suspicion in the villages; but this time the cause was different. One day we received a delegation of Nogah's Iranian and Iraqi Kurds, who had the following to say: "Gentlemen, you brought us here and settled us on the land, which we have been working now for over two years. It is good soil and blessed with crops, but our profits go to others. The ranch development company is exploiting us, paying us a daily wage and selling all our excellent crops to *Tnuva;* together, they are doing good business at our expense."

This was the moment for which we had been praying. We said to them: "We've explained to you that the arrangement with the company was only temporary, until such time as you're able to set up in farming fully on your own. What is it you wish now?"

They answered: "What do we wish? Simply this: give us our plots of land, and we shall cultivate them."

## CHAPTER XXVII

# THE BANK AT NEHORAH

This was the day to which we had been looking forward. For there could be no better proof of the Jew striking roots in his soil, and of his confidence in the labor of his own hands and the blessing of his toil. We started allotting the A lots — of one acre and a quarter close to the homes. The reason we did not immediately allot the settlers their full quota of five acres was because we feared it would be too complicated for them.

Luck was in during their first independent season. Good prices were fetched by the vegetables grown in lot A. Some ready cash started flowing into the settlers' pockets. We came to the conclusion that the time was ripe for opening a bank at the Nehorah rural center.

Calling in the Workers' Bank people from Ashkelon, we asked them to open their first branch in the Lakhish Region. We allotted them a room and a half in a communal building in the center of Nehorah, fitting bars to the windows as befits a bank. One fine day the bank people arrived, bringing with them a counter with a cashier's cage, a desk, a large iron safe and a sign-board with big and clear letters — "Workers' Bank Ltd., Nehorah Branch."

At first the bank stood empty, no one looked in. The settlers had indeed heard of the bank, and some even passed it, on their way back from work in their carts. Others, coming into the clinic or the tractor station at Nehorah, were more daring, peeping in and exchanging a few words with the bored clerk, then departing just as they came.

The bank was paralyzed. Nobody deposited money. No account was opened.

The instructors commenced a persuasion campaign among the inhabitants of Otzem and Nogah. They spoke of the safety of depositing savings in a bank. Not only was a person spared the worry day and night over money hidden away — in jars, deep in the ground or under a mattress — once it was deposited in the bank's armor-plated safe; not only that, but the money deposited in the bank bore fruit by way of interest. And another good reason for opening a bank account: one did not have to carry wads of notes and burdensome coins in one's pockets; instead, there was the check book, by means of which one could effect any manner of payment.

The people listened attentively, asking all sorts of questions, but they kept away from the bank. It was obvious that unless we persuaded the head of one of the large families to open a bank account, the others would not budge.

We addressed ourselves to Yehuda Sharifi, the head of one of the Iranian Kurdish clans at Nogah, and after much persuasion, he agreed to open a current account in the bank with a small deposit.

Thereafter, the bank began to show signs of life. Small accounts were opened, albeit at a slow pace; the bank clerk awoke from slumber; but a credibility gap was still evident between the bank and the settlers.

One day one of the settlers — who had only the week before opened a savings account, depositing one hundred pounds — walked into the bank and, going up to the cashier's cage, asked the clerk: "Tell me, friend, how much money do I have here?"

Checking the man's passbook, the clerk replied : "One hundred pounds, sir."

"And where are my hundred pounds? Where are they located?"

"Over there, in the large safe in the corner," replied the clerk.

"Are you positive all the hundred are here? Maybe you've taken them to Ashkelon or to Tel Aviv, and they're no longer available?"

"You may rest assured, sir," the clerk responded, "that all the money is here. I've already told you so."

"So you've told me, so what! Can't I inquire about my money? It's not your money," the settler was annoyed by then, "and now I want to see the money with my own eyes!" casting an angry glance at the clerk.

"How do you mean, see?" asked the clerk. "Do you wish to withdraw any amount from your account?"

"No," answered the settler with growing suspicion, "I want to see my money and count it."

"But I can't do such a thing," the bewildered clerk said, "unless you make out a check, and then I'll give you money for it."

"O.K.," said the settler, "I write a check." Taking out a check book, he said to the clerk: "Write me a check, and I'll sign."

"For what amount should I make out the check?"

"For a hundred pounds."

"What? Do you want to close the account?"

"I don't want to close the account. I want to count all the money. I don't trust you people."

The clerk made out a check for a hundred pounds, and the settler signed. Taking a packet of notes out of the safe, the clerk counted a hundred pounds and handed the money to the settler. The latter took the notes, and sitting down at the desk in the corner, counted the money over and over. Then with an apologetic smile he returned the money to the clerk, saying: "You're a young man, and haven't yet had a taste of what it means to *earn* money, taking your bread from the soil. But I see that you and your bank are honest folk. Here, take my hundred pounds back and put them in the safe, locking it well."

He added: — "Hold on a minute, here's another ten pounds I earned this week; put them, too, in the safe."

## CHAPTER XXVIII

# A TOWN'S BEGINNINGS

If we had to have recourse to a committee of ten experts to determine the location of a small rural settlement, how much more serious was the question of determining the location of a new town — the capital of the Lakhish Region. We wanted to establish it somewhere in the center of the Region, and the natural choice was near the Plugot (Faluja) junction — the important crossroads, halfway on the Tel Aviv-Beersheba road and on the Ashkelon-Beit Govrin road leading to Jerusalem; in other words, on both the north-south and the east-west axes. An enlarged siting committee set out for the area.

A country town was then in fashion among town planners. A sort of pastoral town with a few thousand inhabitants, each sitting beneath his vine and beneath his fig tree — in single-storey houses with attached quarter- or half-acre plots on which they could grow a few tomatoes or onions and raise some chickens.

This approach was contested by a simpler one, which visualized a town like all towns — based mainly on new industries, with multi-storeyed houses.

When planning Kiryat-Gat, we arrived at a compromise. The first residential quarter would be built according to a semi-rural formula, future quarters being built on an urban formula, having multi-storeyed houses. Industry would be based on the farm produce of the neighboring villages and on local labor. Other industries would also be introduced if the town grew.

In determining the town's exact location, we also took into account the railway track which was then being laid between Lod and Beersheba. We knew that the vicinity of a railway track would be important for industrial development, and the planned industrial zone was therefore brought right up to the railway line.

Once the general contour plan was determined for Kiryat-Gat, which we visualized as a small town numbering some ten thousand inhabitants, we commenced detailed planning for the first neighborhood. Isaac Chizick, with a background of security and administration activity during the British Mandate, was appointed my deputy for matters relating to the town. Within a few weeks, Isaac, his secretary, and his assistant moved into two tiny gray shacks put up on a small hill, on the proposed site of the entrance to the future town.

That was the beginning of Kiryat-Gat.

A short while later, we started laying the foundations for the first residential quarter.

The first subject to occupy our attention was security; who was to protect the town's first settlers? We decided it was essential that the first structure be a police station worthy of the name, which would also serve the future neighborhood settlements, but would in the first instance be a sort of fortress for the inhabitants of the town itself.

We persuaded National Police Headquarters to construct speedily a proper building, along the design of the Teggart Fortresses. It was our wish that the arrival of the first settlers should coincide with that of a dozen policemen with their families, who would occupy the police building and thus be there to greet the first comers.

These plans were submitted for approval to the directorate of the Settlement Department. One of the conferences at which the planning of Kiryat-Gat was discussed was also attended by the legendary Avraham Harzfeld, the "Father of Settlement." At sight of the police building plan, Harzfeld exclaimed: "What are you doing? Are you going to start the first Jewish town in the Lakhish Region with a 'gendarmerie' building? Why not commence by erecting a cultural hall, a community center or the *Histadrut* building? Why just the 'gendarmerie'?"

I explained that the principal function of the "gendarmerie" at Kiryat-Gat resembled that of the "tower and stockade" in the agricultural settlement. That set Harzfeld's mind at rest.

Our next concern was to bring in industry. Cotton growing had just begun in Israel in those days. Sam Hamburg, a wealthy farmer from the United States who had studied in Palestine in his youth, arrived in Israel with the tidings : "The country's soil is suitable for cotton growing." The farmers entertained doubts at first, but old Sam swept them away with his enthusiasm; and when we tackled the agricultural planning of Lakhish, we already included a few thousand acres of cotton in the crop rotation plans. The same was done by the Negev Region planners and the farmers of the Ashkelon coast settlements.

The question of a carding machine then confronted us. There were no carding machines as yet in Israel, the first two were then en route from America. One was earmarked for the north, for Havat-Shmuel in the Beisan Valley, which Sam Hamburg had set up as Israel's first cotton-growing ranch. There were many competitors for the carding machine designated for the south: Beersheba, Ofakim and Ashkelon. And now we, too, went to Pinhas Sapir, then Minister of Commerce and Industry, and to Haim Gvati, who was

Director-General of the Ministry of Agriculture, with our claim: "Kiryat-Gat is entitled to receive the south's carding machine. The plans call for the Lakhish Region to be a major cotton-growing center in the south; we have a new town in Lakhish, and it has no industry."

It appeared to us at first that we would win the "battle of the carding machine," since the Beersheba and Ashkelon settlers' arguments were weaker than ours. However, they then played their strong card:

"It's now May, and the carding machines are already on board the ship making its way to Israel," they said. "It's essential that the machines be installed immediately upon the ship's arrival, otherwise they won't be able to go into operation in time to process the coming season's crop, that is, this November or December. Our towns are already established, with industrial centers, with roads, water and electricity; construction of the necessary structure can begin immediately, and it can be ready in time to handle the first cotton crops.

"And you? Your entire town, after all, is still on the drawing board. You have no electricity, no roads, no water; how are you going to set up the carding machine within the space of five months?"

Taking a deep breath, we said: "We undertake the responsibility to have the carding machine in Kiryat-Gat linked within a few months to the water, electric power and road networks."

Sapir and Gvati supported us, deciding that the carding machine should be put up in Kiryat-Gat.

Nahum Gershonovitz was appointed special assistant to Isaac Chizick for setting up the carding machine. I issued instructions to the engineers to make every effort, draw on all resources of design and resourcefulness, improvise as much as possible — just so long as we met the timetable.

We decided immediately to take a risk, and not construct

the entire structure for the carding machine. We knew we would not be able to make it in time. Instead, we decided to lay concrete foundations only, on which we would place the machinery.

The work was conducted at a hectic pace. A narrow limestone road was laid to the "industrial zone." We poured the concrete surfaces, connected a provisional water pipeline to Moshav Uzzah, and drew a provisional electricity line from somewhere. The machines were installed at great speed on the surfaces in the open.

The first day of cotton picking in the fields of the south, we made a test run of the machines, standing under the open sky.

We had beaten the deadline.

When the steel teeth carded the first cotton fibers, we felt like the person who — exerting his last ounce of strength to catch the last bus — takes a flying leap at the very last moment, lands on the step and, ascending into the bus, thanks the driver and sinks bone-weary into the seat.

## CHAPTER XXIX

# ASTARTE OF TEL-GAT

Even before the first families arrived at Kiryat-Gat, we were concerned about their livelihood. Agriculture was out of the question, since we were dealing with a town, and, anyway, there was hardly enough agricultural work for the village settlers.

It was, of course, possible to employ some of the job-seekers as unskilled laborers, in construction work. But what about the others? We gave thought to the matter, and suddenly had a bright idea:

Here was Kiryat-Gat being built next to an ancient *tel*, known as Tel-Gat. It had not been touched by archaeologists, and was by way of being a virgin *tel*. Who knew what it concealed? And was it not time for its secrets to be unveiled?

We contacted Professor Yeivin, Head of the Government Department of Antiquities.

Yes, said the Professor, provided we supplied a budget for an archaeological expedition, and if we gave implements, the necessary structures and equipment, he would be prepared to work the *tel*, employing daily about a hundred laborers — provided, of course, we paid their wages.

Eureka! We had found creative, even educational, work. But where was the budget coming from?

I called on Eshkol in Jerusalem. He was then Minister of Finance, as well as Head of the Settlement Department.

"What new trouble are you bringing me this time?" Eshkol asked pleasantly, when receiving me in his office. "Start from the end," he would say.

I related briefly what it was all about, glorifying and exalting the historical and cultural importance of Tel-Gat. I contended that the place would turn into a tourist site, becoming a source of revenue for the State and for Kiryat-Gat. While still at home I had perused the Bible in anticipation of my confrontation with Eshkol, and in order to impress him, produced a relevant quotation.

"It says here, that there were golden rats and mice in the Philistine temples. Who knows but that we may uncover a hoard of such rats and mice in the dig, and that will put paid to all your worries; admittedly, rats and mice are not the most respectable of things, but gold purifies all, as it has been said. It is likely to purify our foreign currency deficit."

Eshkol cast a fatherly glance at me over his spectacles. I knew he was hunting for a counter-phrase: "Do you recall," he said, "when the young David fled from Saul, and found refuge with Achish, king of Gath?"

"Yes," I answered with some trepidation.

"And also that, coming before the king of Gath, David feigned madness, and the king of Gath said: 'Do I lack madmen?' I'm neither king nor son of a king, but let me tell you, I don't lack madmen coming to me every day with all sorts of ideas of how to save the country and its economy."

Sensing that this was not an absolute "no," I worked a bit more on Eshkol, until he granted my request and added some 20,000 pounds to our budget, for excavation at Tel-Gat.

We immediately informed Prof. Yeivin, who started organizing an archaeological expedition. We, for our part, hurriedly put up a few prefabricated huts on the slopes of the *tel*, brought in beds and mattresses, and constructed some shower stalls and toilets.

A few weeks later Prof. Yeivin turned up with some of his people, to examine the little camp we had put up for them. Neither the huts nor the other fixtures satisfied him.

He was particularly indignant about the toilets, which had been hastily installed in tiny cabins. "What kind of seat is this?" the heavyset Professor asked me, pointing to the wooden seat with the round hole in its center. "What caliber did you pick on here? This is for babies!"

We enlarged the caliber as required.

Two days following the arrival of the first settlers at Kiryat-Gat, the archaeological expedition also made its appearance, taking up residence in the camp. We sent out about a hundred people to work, all of them from Morocco and the Atlas Range, equipping them with the requisite tools — hoes, spades and baskets made of old tires. The archaeologists instructed them how to dig and uncover the *tel*. At first, our people did not grasp what was required of them: here they are taken out to the *tel* and told to dig very gently, filling the baskets with earth. Why and wherefore was this? What was going on here in this strange new land?

We explained to them, as best as we could, that in this *tel* there had once been a Philistine town, and maybe even Goliath had lived there. The name Goliath made an impression. Happily, the first layer to be uncovered contained a burial ground a thousand years old or so. The laborers, under the archaeologists' supervision, collected the bones carefully in their baskets. When one of them — they were mostly bearded and galabia clad — would find a particularly large bone, he would sing out: "Here's Goliath's leg; here's Goliath's arm."

One other phenomenon had them perplexed: Orah. She was a member of the expedition, and she had a trim, magnificent body and very lovely, tanned long legs. Her only real apparel was a huge wide-brimmed hat, protecting her from sun rays; for the rest — extremely short pants, practically a bikini, and a very thin blouse.

The Atlas women are covered from head to foot, and our folk looked upon Orah as though she were a creature from another world, beyond their reach. She and all this archaeology business too! Days and weeks passed, and Orah's tan deepened, until she shone like burnished bronze.

Digging deeper, the laborers started uncovering shards of jars and other vessels. Layer after layer was discovered. The archaeologists were excited over every find, bringing in healers to mend and stick together bits of jars and photographers and surveyors to reconstruct the *tel*, and there was much rejoicing. The laborers were also cheerful, for they received their wages on time.

Meanwhile, the cotton carding machine was up, and there were indications of new sources of employment in Kiryat-Gat.

One day the archaeologists showed me a small figurine of Astarte, the Canaanite goddess of fertility, which had just been unearthed. It was a lovely statuette of a naked female, full-breasted and wide-hipped, made of red clay. I requested permission to take it and show it to Eshkol. The archaeologists reluctantly granted my request; they wrapped the lady in cotton-wool and placed her in a small cardboard carton.

On my next visit to Jerusalem, I took the little Astarte with me. At the height of a budgetary discussion held in Eshkol's room, in the presence of a number of participants, I proudly removed Astarte from the carton, placed her on the desk in front of Eshkol, and said to him: "There, see what lovely finds there are at Tel-Gat."

Glancing at the little goddess with amused surprise, Eshkol picked her up in his large fist, drew her closer, and, tapping lightly with his fingers on her buttocks and breasts, said: "What an impudent Canaanite female. Not enough for her what she extorted from the Israelites thousands of years ago, when they whored after her on every hillock and under every luxuriant tree; she has now succeeded in extorting another 20,000 pounds, in hard cash, from the Treasury of the State of Israel."

## CHAPTER XXX

# THE SHORT CUT

After the first cotton carding machine had been set up in Kiryat-Gat, we addressed ourselves to another agricultural crop requiring special attention: groundnuts. Already in the first year the Region's settlements had a large yield of ground-nuts, which had to be sorted and packed. We put up a large sorting and packing shed in Kiryat-Gat, thus providing a few more thousand industrial workdays, mainly for women.

At the same time we looked for investors to establish a permanent industry, which would not be so dependent on seasonal work as were agricultural by-products.

Textiles — a Jewish industry by origin — was then making its first steps in Israel. That industry provides rela-tively high employment and our search for investors natu-rally veered in that direction.

It was thus that we hit upon the idea of establishing a cotton yarn spinning mill at Kiryat-Gat. There were, after all, ex-tensive cotton fields in the Region, and even a carding ma-chine was already on hand. So why not proceed to the logical conclusion?

We requested the Ministry of Commerce and Industry and other agencies to find us investors for a spinning mill. It was not long before they introduced us to a local person who had connections with a group of Jewish investors in America, who were interested in establishing a spinning mill in Israel.

Negotiations had been conducted throughout by Isaac Chizick, my deputy for Kiryat-Gat. I saw the Israeli investor only a few times.

When the Americans arrived at Kiryat-Gat, Chizick gave them a royal welcome. For it was no little matter: the first investors "after two thousand years"!

Owing to other engagements, I was unable to attend. Impressed by both the pioneering vision and the favorable financial terms, the investors departed satisfied from Kiryat-Gat.

A few days later, I received a phone call from the attorney acting for the group of American investors. He had a request to put to me, the Director of the Region: would I be kind enough to come to his office in Tel Aviv on a certain date, to meet these investors?

"You understand, Mr. Director," said the courteous attorney, a Hungarian Jew, "these investors have heard a lot about you, and would like to meet you and talk to you, since you are sort of governor of the entire project. I, too, have heard a lot about you, and should also be pleased to make your personal acquaintance."

I said that I was honored, and we arranged a meeting at his office at 10 a.m. the following Monday.

Harnessing my jeep station-wagon, I arrived at ten sharp for the meeting at the attorney's comfortable Tel Aviv office. I was then in my early thirties, thin and lean of figure clad in an open-neck khaki shirt and desert shoes.

A pretty receptionist answered my ring at the bell. I told her I had an appointment with the attorney, and she led me into his room. It was a large room, with about half-a-dozen people reclining in cozy armchairs.

I stood like a beggar in the doorway, with nobody announcing my arrival. After an uncomfortable moment's silence, an elderly person stood up, his white hair beautifully

combed and arranged like a silver circlet. Stepping up to
me, he looked me over and asked: "Has your car had a
breakdown? Will Mr. Eliav be arriving for the meeting?"

I told him: "I'm Eliav; I understand I have the honor of
addressing attorney..." The man blanched and blushed, but
soon took hold of himself. "Excuse me, Mr. Director," he
said, bowing and shaking my hand.

Turning around, he drew me into the center of the room.
Warmly embracing my shoulders, he addressed the Americans:

"My dear gentlemen," he announced in a resonant voice,
"look what young, simple-mannered and democratic directors
we have in Israel! I have the honor, the pleasure and the
great privilege to present to you the Director of the Lakhish
Region." He then proceeded to praise me, detailing all
the good deeds I had done from kindergarten onward.

Shaking the Americans' hands, I sank into one of the
leather armchairs, and the conversation flowed smoothly.
Glasses of champagne were handed round. We toasted the
spinning mill, American Jewry, Israel, Kiryat-Gat, the Ministry of Commerce and Industry, the attorney's office, and
so on. A photographer appeared from somewhere and com-
memorated the festive occasion.

Not many months later, the "Lakhish Spinning Mill"
was established at Kiryat-Gat.

The next stage was to expand the textile industry at
Kiryat-Gat. With the existence of cotton, a carding machine
and a spinning mill, why not go into clothing manufacture?
After all, ready-made clothing was also a Jewish industry.

One of the magnates of Latin America's textile industry
arrived in Israel at the time. Israel Pollak came from a family
who had had a textile plant in Romania for generations, and
was a Zionist from youth. During the Second World War he
immigrated to Siberia, and both there and in Central Asia he

set up textile industries for the Soviets. After the war he landed in Chile, where he established, together with his brothers, giant enterprises in the spheres of weaving, spinning, knitting and the various branches of ready-made clothing. He employed thousands of workers, and everything he touched was a success.

Pinhas Sapir, who was then Minister of Commerce and Industry, met him during a visit to Chile and, playing on his warm Jewish heart, talked him into immigrating and investing in Israel. Now that Pollak had arrived in Israel, Sapir wished to interest him in setting up his enterprises at Kiryat-Gat. There was a question here of a *real* industry employing hundreds of workers, and involving an entrepreneur of the first order. Knowing that if he was successful, it would mean a major revolution for the town, we followed Sapir's actions with bated breath.

During his first talk with Pollak, Sapir painted a glowing picture of Israel in general and the Lakhish Region in particular, with special emphasis on Kiryat-Gat: a good climate, water in abundance, a plentitude of land for development, a high-tension power cable practically under one's nose, and no lack of workhands, whether male or female. Thus he went on enumerating Kiryat-Gat's advantages.

Pollak listened attentively, but when he saw Kiryat-Gat's geographical location on the map, he said to Sapir: "And what about the distance? Kiryat-Gat is quite far from the center of the country, from Tel Aviv and its environs. This will increase production and marketing costs; furthermore, there are no banks in the place. Can't you suggest a place nearer Tel Aviv?"

"Nearer?" Sapir exclaimed. "What's the matter with you, Mr. Pollak? It's a distance of no account, no distance whatever; one two, and you're in Kiryat-Gat."

"One two?" queried Pollak. "How long does the trip take

from Tel Aviv to Kiryat-Gat?"

"Ah, really, what's there to talk about?" answered Sapir. "In less than an hour you're in Kiryat-Gat."

"Less than an hour? And I was told at least an hour-and-a-half," said Pollak, surprised.

"You know what?" said Sapir. "Tomorrow morning I'll pick you up at the Dan Hotel and take you to Kiryat-Gat."

"Good," said Pollak, "I shall be greatly honored. What time shall I expect you in the morning?"

Sapir: "What time? Say five o'clock."

"Five a.m.?" Pollak was astounded, but made no further comment.

At five sharp the following morning, Sapir arrived at the hotel in his official car. Pollak was already waiting, and they set off.

The driver, who had earlier been given concise instructions, stepped on the gas, and the speedometer soon registered 60 m.p.h. The car sped through the empty streets, and the moment they were out of Tel Aviv the driver increased the speed to 75 m.p.h. and more.

En route, Sapir and Pollak made small talk. At 5.55 a.m. the car stopped near Kiryat-Gat's future industrial area.

Sapir and Pollak climbed to the top of a nearby hillock. The sun had just risen, the landscape was beautiful, and Sapir's words were sweeter than honey.

They stayed a few minutes on the hillock, and by about 7 a.m. were already back at the Tel Aviv hotel.

Pollak, wise and clever, fully grasped the purpose of the entire exercise, including the choice of the early morning hours. "But I thought to myself," as he subsequently told me, "if an Israeli Cabinet Minister is prepared to rise at four o'clock in the morning, to persuade me in this fashion to establish an enterprise at Kiryat-Gat, who am I to contradict him."

Thus was established at Kiryat-Gat the large Polgat complex, which eventually set up the Ouman and Begir department store. It is still going strong.

We wanted Kiryat-Gat to have not only textile and agricultural enterprises; we also wanted just plain folk engaging in ordinary industries to settle there. Locksmith, carpentry and tinsmith industries were established in the town in due course, as well as a number of plastics factories, and the place gradually took in more and more enterprises and small projects.

Among these was also a candle factory. Some years later when I served as Deputy Minister of Commerce and Industry, and was in charge of industrialization of development districts, I visited Kiryat-Gat and her industries. I included the candle factory in my itinerary. I was received by the two owners, pleasant, orthodox Jews, skullcap on head and four-fringed garment peeping beneath jacket.

They took me around from room to room and from one machine to the next. Seated at the first machine were girls. The machine stamped the strips of liquescent wax with a monotonous beat, and slender white candles were pressed through tubes.

"These are Sabbath candles," the owners explained, "they are packed by the girls in special packets." I noticed the closed packets with their legend in large type: "Sabbath candles."

At another machine we observed thicker strips of wax being cut by an automatically operated blade and then poured into tiny tin cups.

"These are memorial candles," my hosts told me.

The third machine made *Havdalah* tapers.

A fourth machine, working feverishly, produced hundreds of slender colored candles.

"These are *Hanukah* candles. Those packets there, with

their Shields of David, are for export to the Jewish communities in America and Europe."

I evinced great interest in the factory's export potential. Seated in the small office on the second floor, we discussed problems and possibilities, the shortage of operating capital, and prospects of growth and expansion.

There was a miniature display of all the products in a glass-fronted case: Sabbath candles, memorial candles, *Hanukah* candles, *Havdalah* tapers, *Yom Kippur* candles and the like. At the very end of the line there was another candle, but covered.

"And what candle is that?" I inquired curiously.

With a sheepish grin at each other, they removed the cover.

It was quite a candle! The Venus de Milo must have served as its model, only this one had arms and hands.

Looking her over, I asked incredulously: "What's this swinger doing among the other candles?"

"Ah, well," said the candlemakers, "there's a great demand for such candles in cabarets and discotheques."

## CHAPTER XXXI

# PROTECTZIA

After Moshav Shahar had been settled by North African immigrants, many of them from Tangier, one section remained unoccupied.

One day my Cochinese friends from Nevatim visited me in Ashkelon, among them my old aide, Nissim Eliahu.

They had a big favor to ask: more families were due from India, some of them relatives of Eliahu Nissim and his friends, but Moshav Nevatim was already fully settled. Eliahu himself was about to get married. In short, they were very keen on being absorbed in one of the Region's settlements.

There were too few of them to populate an entire *moshav*, and that gave me the idea of putting them in Moshav Shahar.

When I brought my proposal before the team, there was quite some opposition to the idea. Past experience had shown that a hasty and artificial fusion of communities was an inexhaustible source of friction. I was adamant, and knowing my special regard for the Cochin immigrants, my colleagues finally accepted my idea.

About a dozen families of Cochin immigrants thus arrived at Shahar, among them the Nissim Eliahus, Eliahu Eliahus, Bezalel Eliahus, Meir Eliahus ...

The first days and months at Shahar were very difficult for them. For all their kindheartedness and the special brand of tranquillity typical of them, their integration with the North African immigrants was not easy. Their absorption in farm life was also difficult, but the great pace of development which swept across all the Lakhish settlements infected them as well.

The years passed. Shahar's Cochin immigrants stuck to the land. In the beginning, they lived sparsely. Like all other Lakhish settlers, they engaged in field crops, such as sugar beet, cotton and assorted vegetables, making a scanty living. I never heard them complain. They would recount their troubles to me in the privacy of their homes; but it was invariably with an apologetic smile. Their homes were always clean and carefully tended, as were their handsome children.

In those days, there were a few flowering shrubs growing outside the stark and shadeless houses. They brought with them from Cochin a flair for growing exotic plants. They brought with them all kinds of seeds and strange spices, and tried to grow them next to their homes. Thus Shahar's Cochinese settlers were the pioneers in the immigrants *moshavim* in growing flowers.

Upon my return to Israel after a few years' service at the Israeli Embassy in Moscow, the Shahar settlers gave me a very hearty reception at one of the village homes. At the end of the evening, Bezalel's family handed me a bouquet of gladioli, and Bezalel stated proudly that these were the first crop of the Lakhish Region's flowers. "We get fairly good prices for them from the Tel Aviv shops. They're also talking of exporting flowers."

Bezalel — who already then, at the beginning of the 'sixties, displayed an uncommon flair for flower growing — was sent by the Ministry of Agriculture to a number of European countries to study this field, particularly hothouse roses.

211

Some more years passed, and again I went abroad. I was sent by Israel to coordinate agricultural development work in Iran's Kazvin Region, as well as in other countries.

When I came home, not so long ago, I again visited the Lakhish settlements, calling at Shahar as well. My surprise may well be imagined. I was aware that the flower growing branch in Israel was expanding at a terrific rate; I also knew that hothouses, both of plastic and glass, were in ever-growing use, and that the branch was becoming more and more industrialized. But I had no idea that Bezalel and some of his friends at Shahar had assumed a leading role in the field, and that they ranked in Israel's "national team" where flower growing was concerned.

Bezalel invited me to tour his project. Large hothouses had been erected near his and other homes at the *moshav,* Bezalel's hothouse occupying more than half-an-acre; a veritable industrial enterprise. Installed in the big hothouse was a complex electronic system of heating and chilling, careful irrigation and scientific fertilizing; the entire hothouse was a blooming garden of roses and rosebuds in all colors of the rainbow.

Next to the hothouse was a sorting and packing shed. It was twilight; the lads and lasses, as well as the boys and girls, returning from field and school, were all engaged in sorting and packing the roses.

A capacious van was parked outside Bezalel's home. "Everything is for export," Bezalel told me. "We have a direct phone line to the 'Agrexco' office at Lod Airport, and they communicate by telex with the European flower wholesalers. Here in Shahar we have to know day by day what type and color of roses will be required the next day by housewives and loving couples in Zurich, Bonn, Geneva and London.

"The packed and chilled flowers are loaded on to the van,

which sets out for Lod in the evening. As you can see, every parcel of roses has a tag attached, on which is marked the foreign destination. The flowers are airfreighted by El Al during the night or early the following morning, and by noon they are already in Europe's flower shops."

As if by the way, Bezalel told me that the capital investmen in his hothouses amounted to about a quarter-of-a-million pounds. The lion's share, naturally, came from loans granted him at the recommendation of the Ministry of Agriculture.

Bezalel also told me that Yehezkel Elias had spent about two years in Liberia, where he had served as an instructor on behalf of the International Cooperation Department of Israel's Ministry for Foreign Affairs, and had assisted in the training of Liberian farmers in modern farming methods.

I left Bezalel's home with the settlers' greetings ringing in my ears, and a huge bouquet of roses in my car.

Some time later, at Kfar Vitkin, I met the instructors who had worked with me in 1954 at Nevatim in the Negev, when we absorbed the early Cochin immigrants. Naftali and Ami ("Keuke") nowadays own flourishing farms at Kfar Vitkin, and have wives and children. The only reminder of those good old days is Naftali's huge mustache.

Orah and Uzzi were there as well. They had built their home at Moshav Nir-Banim, on the northern fringes of Lakhish, a flourishing *moshav* of old-timers. They owned a large dairy farm, and had three lovely children.

We chatted about current affairs, and then, as is the case among farmers and *moshav* people, the conversation turned to farming and its various branches, eventually coming round to the flower-growing branch. I heard from them that many of Nevatim's settlers were nowadays good flower growers, but not one of them came anywhere near being as successful as Bezalel at Shahar.

"Oh," Orah exclaimed with a glint in her eyes, "Lyova

can arrange some *protectzia* for us with Bezalel."

"What is it?" I asked. "What kind of *protectzia*?"

"It's like this," Orah replied. "Some of Nir-Banim's settlers, ourselves included, are engaged in flower growing, gladioli in particular. Now, your Bezalel knows the secret of growing a wonderful type of dwarf-gladioli. The demand for dwarf-gladioli in Europe is something extraordinary. Could you ask Bezalel to let us into his secret?"

But, when the wheel has turned full circle like that, whose secret is it, really?

# GLOSSARY

Akiva, Rabbi — leading rabbinical authority and teacher in Judea in 2nd Century C.E., who gave motivating force to Simon Bar-Kochba's revolt against the Romans (132-135 C.E.)

Aliyah (Second and Third) — immigration (to Palestine / Israel) — two of the waves of immigration to Palestine (1904-1914; 1919-1923).

Arava — the part of the Negev (q.v.) running along Israel's border with Jordan, from the Dead Sea to the Gulf of Eilat.

A.T.S. — Auxiliary Territorial Service (one of the women's services in the British Forces).

Bar Mitzvah — the age at which a Jewish boy assumes religious obligation (13 years of age).

Conquest of Jewish labor — the efforts of early Jewish immigrants to do work not generally done in the Diaspora, or done by non-Jews in Palestine.

Deganya Bet — one of the oldest kibbutzim, south of the Sea of Galilee (Bet = letter B, second).

Egged — largest bus cooperative in Israel, serving most of country.

Ein Farah — source of *Wadi Kelt* (q.v.), north-east of Jerusalem.

Falafel — ground chick peas with spices, rolled into

Fedayeen — Arab terrorists, bands of whom were formed by the Egyptians from among the Palestine Arab refugees in the Gaza Strip.

Field Force — infantry units of the *Haganah* (q.v.).

Gadna — Israel's youth battalions, para-military.

Geulah — Hebrew feminine name (lit. "redemption").

Gosplan — Soviet State Planning Authority.

Haftarah — chapter from the Prophets read in synagogue after the Portion from the Pentateuch.

Haganah — voluntary Jewish self-defense organization established in Palestine, especially against Arab attacks, during the British Mandate; forerunner of the Israel Defense Forces.

Hakham — sage (Rabbi) among Oriental Jewish communities.

Halutz(im) — Jewish pioneer immigrant in Palestine. (plural)

Halutziut — pioneering.

Hamashbir — Israel's largest cooperative wholesale stores.

Hamashbir La'Oleh — cooperative wholesale stores for the immigrant.

Hapoel Hamizrachi — religious organization (principal constituent in Israel of National Religious Party).

Hashomer — Jewish self-defense organization founded in Palestine in 1905.

Hassidim — members of Hassidic (orthodox) sect.

Hatzav — squill.

Havdalah — benediction at the conclusion of Sabbaths and festivals.

Heder — elementary Jewish religious school.

Histadrut — Israel General Federation of Labor.

Kaffia — traditional Arab headdress.

Keren Hayesod — United Jewish Appeal (England — Joint Palestine Appeal).

216

## Glossary

| | | |
|---|---|---|
| Keren Kayemeth | — | Jewish National Fund. |
| Kfar | — | village. |
| Khan | — | old Arab or Turkish inn. |
| Kiryat | — | town, city (Kiryat... — town of...). |
| Kolkhoz | — | collective farm in Soviet Union. |
| Kupat Holim | — | health insurance plan [the reference is to the Sick Fund of the Histadrut (q.v.)]. |
| Large Crater | — | one of three crater-like geological depressions in the Negev (q.v.) [Small Crater, Large Crater and Ramon Crater]. |
| Lod | — | Lydda. |
| Ma'abara(ot) | — | transit camp built in Israel for immigrants. (plural) |
| Mammeliga | — | ground corn usually in consistency of porridge. |
| Mar Saba | — | ancient monastery in Judean Desert, between Jerusalem and Dead Sea. |
| Mekoroth | — | Israel's National Water Company. |
| Mes'ha | — | original Arab name of Kfar Tabor in Lower Galilee [birthplace of Yigal Allon]. |
| Metzudat | — | Fortress of [metzuda — fortress]. |
| Mezuza(ot) | — | parchment scroll containing Deut. 6:4-9 and 11:13-21 affixed to the doorpost of Jewish homes in a wooden or metal case. (plural) |
| Mishnah | — | collection of Jewish oral laws. |
| Moshav(im) | — | smallholders' cooperative settlement. (plural) |
| Moshavnik | — | member of *moshav* (q.v.). |
| Mossad | — | Institute. |
| N.A.A.F.I. | — | Navy, Army and Air Force Institute [British equivalent of PX]. |
| Nahal | — | Israel's Pioneer Youth Corps. |
| Narodnik(s) | — | Russian pre-Revolutionary Populists, of the late 19th and early 20th centuries, land |

217

reform being one of their principal aims.

Negev — Israel's arid southland.

Nudnik — Yiddish slang for a bore, nuisance, pest, nagger.

Oron — phosphate works in Negev (q.v.), near rim of Large Crater (q.v.).

Palmah — striking force of *Haganah* (q.v.).

Palmah-Yam — naval arm of *Palmah* (q.v.).

Patinkin, Prof. Dan — Professor of Economics at Hebrew University of Jerusalem and an Economic Adviser to the Israeli Government.

Piaster — a monetary unit and coin of Turkey and Egypt, current in Palestine and in early days of State of Israel.

Protectzia — favoritism, "influence," "pull" [slang].

Rehov — street.

Sabra — Israeli-born [lit. prickly fruit of cactus], referring metaphorically to prickly exterior and tender heart.

Schnor — begging, fund-raising [Yiddish slang].

Sdom — location of Dead Sea Potash Works at south-western corner of Dead Sea [Biblical Sodom, presumed location of which is further to the south].

Shalom — Hebrew greeting, on arrival and departure [lit. peace].

Shomron — lit. Samaria; the reference in this context is to that part of the District of Samaria which is in Israel proper.

Solel Boneh — construction company founded by the Histadrut (q.v.).

Sovkhoz — state-owned farm in the U.S.S.R. paying wages to the workers [compare *kolkhoz*].

Sovnorkhoz — Soviet economic administrative unit.

| | | |
|---|---|---|
| Taanach | — | Taanach by the waters of Megiddo [Judges 5, 19]. |
| Taref | — | non-kosher (food), forbidden (food, vessels). |
| Teggart Fortresses | — | British Mandatory police fortresses, named after designer, Sir Charles Teggart. |
| Tel | — | archaeological mound. |
| Tikvah | — | Jewish feminine name (lit. "hope"). |
| Timna | — | Timna Copper Works in the Negev (q.v.). |
| Tnuva | — | Israel's largest workers' cooperative for marketing and distributing farm produce. |
| Tower and stockade | — | system of building Jewish settlements in Mandatory Palestine, instituted during 1936-39 riots. |
| Trumpeldor, Joseph | — | Jewish officer in Russian Tzarist Army, who became a legendary hero. Organized, with Ze'ev Jabotinsky, the Jewish Legion. Died defending Tel-Hai in Upper Galilee against marauding Arabs, where his grave has become a national shrine. |
| Tzrifim | — | former British Army encampment of Sarafand, on the outskirts of Ramlah in central Israel. |
| Um Junni | — | former swampland in Jordan Valley, south of Sea of Galilee, where Deganya is located. |
| Wadi | — | watercourse that is dry except in rainy season. |
| Wadi Kelt | — | runs from *Ein Farah* (q.v.) to River Jordan, north of Dead Sea. |
| Wingate, Brigadier Charles Orde | — | as Captain in the British Army in pre-World War II Palestine, trained the *Haganah* (q.v.) night squads in guerrilla tactics and led them in raids against Arab marauders. Subsequently commanded the famous Chindits in Burma. |

Yeshiva — Orthodox Jewish school or college for religious education; Talmudic college.

Yishuv — the Jewish population in Palestine.

Yom Kippur — Day of Atonement.